GCSE
Success

Biology

Ian Honeysett

Contents

A Balanced Diet

Different Food Molecules

All organisms need food to survive. Food is needed for raw materials for growth and to provide energy. We take in our food ready made as complicated organic molecules. These food molecules can be placed into seven main groups (see table below).

A **balanced diet** needs the correct amounts of each of the types of food molecules.

Different people may have slightly different diets due to a number of factors. These include:
- their **age** and **gender**
- their level of **activity**
- whether they are **vegetarians** or **vegans**
- whether they have any food **allergies**.

Food type	Made up of	Use in the Body
Water	Hydrogen and oxygen	Prevent dehydration
Minerals	Different elements, e.g. iron	Iron is used to make haemoglobin
Proteins	Long chains of amino acids	Growth and repair
Carbohydrates	Simple sugars, e.g. glucose	Supply or store of energy
Vitamins	Different structures, e.g. vitamin C	Vitamin C prevents scurvy
Fats	Fatty acids and glycerol	Rich store of energy
Fibre	Cellulose	Prevents constipation

💡 Boost Your Memory

You can use this mnemonic to remember the seven types of food molecules: **W**hen **m**y **p**arents **c**ook, **v**egetables **f**eel **f**unny.

This stands for: **w**ater, **m**inerals, **p**roteins, **c**arbohydrates, **v**itamins, **f**ats and **f**ibre.

Alternatively, you can make up your own mnemonic.

✓ Maximise Your Marks

Remember, a balanced diet is the 'correct' amount of each food.

In exams, candidates can lose a mark by saying a balanced diet contains 'enough' food.

Build Your Understanding

Proteins are needed for growth and so it is important to eat the correct amount.

This is called the estimated average requirement (EAR) and can be calculated using the formula:

EAR (g) = 0.6 × body mass (kg)

Too little protein in the diet causes the condition called kwashiorkor. This is common in developing countries due to overpopulation or a natural disaster and lack of money to improve agriculture.

Build Your Understanding

The EAR is an estimate of the mass of protein needed per day based on an average person. The EAR for protein might be affected by factors such as age, pregnancy or breast feeding (lactation).

Some amino acids can only be obtained from the diet and they are called essential amino acids. Although proteins cannot be stored in the body, some amino acids can be converted by the body into other amino acids.

Proteins from meat and fish are called first class proteins. They contain all the essential amino acids that cannot be made by the body. Plant proteins are called second class proteins as they do not contain all the essential amino acids. The amino acid that is in shortest supply is called the limiting amino acid. This will restrict the growth of the person.

If you eat too much fat and carbohydrate, they are stored in the body. Carbohydrates are stored in the liver as glycogen or are converted into fats. There is a limit to how much glycogen the liver can store, but fat storage is not so limited. Fats are stored under the skin and around organs as adipose tissue.

Although proteins are essential for growth and repair, they cannot be stored in the body.

Adipose cells

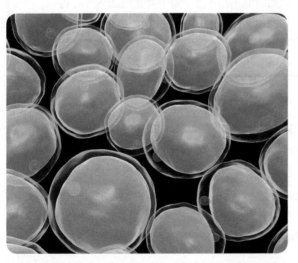

✔ Maximise Your Marks

To get an A* you must be able to analyse data about different types of protein. The Stretch Yourself section will help you to practise this.

Daily Diet

Different foods contain varying amounts of the different food molecules. Therefore, to achieve a balanced diet, people need to eat the correct food in their diet.

The right balance of food intake is approximately:
- Fruit and vegetables – eat at least five a day.
- Bread, other cereals and potatoes – these provide carbohydrates.
- Milk and dairy foods – people should choose lower-fat types and limit themselves to two or three portions a day.
- Foods containing fat, and foods and drink containing sugar – these should be kept to a minimum.
- Meat, fish and alternatives – people should choose two or three lower-fat portions a day.

? Test Yourself

1. What are proteins made of?
2. Tom has a mass of 55 kg. What is his EAR for protein?
3. What is kwashiorkor?
4. Why is it important to have enough fibre in your diet?

★ Stretch Yourself

Look at the table showing details of four foods.

Food	Protein quality rating	Limiting amino acid
Egg	0.98	None
Beef	0.77	None
Wheat	0.62	Lysine
Peas	0.49	Methionine

1. Why do egg and beef have a higher protein quality rating?
2. Why do many diets often involve eating peas with wheat?

Homeostasis 1

Principles of Homeostasis

It is vital that the internal environment of the body is kept constant. This is called **homeostasis**.

The different factors in the body that need to be kept constant include:
• water content
• temperature
• sugar content
• mineral (ion) content.

Many of the mechanisms that are used for homeostasis involve **hormones**. Hormones are chemical messengers that are carried in the blood stream.

They are released by glands and passed to their target organ.

Hormones take longer to have an effect compared to nerve impulses, but their responses usually last longer.

The homeostasis control mechanisms in the body work by **negative feedback**.

This means that **receptors** in the body detect a change in the body.

These changes are then **processed** in the body.

Then **effectors** bring about a response that reverses the change so that the normal level is restored.

This is much like many artificial control systems such as the temperature control in a house.

Receptors, like a thermostat in a room, detect the change

A processor compares the stimulus with a set point, like the temperature on the thermostat dial

An effector produces a response to correct any difference between the new level and the set point, like a radiator being turned up

💡 Boost Your Memory

Many control devices in the home work by negative feedback. Thinking about how they work might help you remember what is meant by negative feedback.

Controlling Blood Sugar Levels

It is vital that the sugar or glucose level of the blood is kept constant:
• If it gets too low then cells will not have enough to use for respiration.
• If it is too high then it may start to pass out of the body in the urine.

Insulin is the hormone that controls the level of glucose in the blood.

Insulin is made in the pancreas. When glucose levels are too high, insulin is released. The insulin acts on the **liver**, causing it to convert excess glucose into **glycogen** for storage.

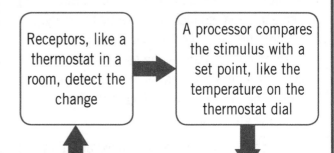

Glucose in the blood — Insulin → Glycogen in the liver

Diabetes

People who cannot control their blood sugar levels have a condition called **diabetes**.

This often causes blood sugar levels to be too high and so glucose passes out of the body in the urine. This can be tested for using testing strips.

There are two types of diabetes:
- **Type 1** diabetes is a genetic disorder and is caused by the pancreas failing to make enough insulin. It is treated with regular insulin injections in order to control the level of glucose in the blood. People also need to control their diet carefully.
- **Type 2** diabetes is caused by the cells of the body failing to respond to insulin and is controlled by making sure that the person does not eat too much carbohydrate in one meal.

People with Type 1 diabetes need to inject themselves with insulin. It is important that they get the dose right, so they have to test their blood to see how much sugar is present.

This will vary depending on:
- When they last had a meal.
- How much exercise they have done recently.

Glucose levels will be high following a meal, and low following exercise.

Insulin injection

Build Your Understanding

Some systems have more than one effector that can work in opposite directions. This means that the response can happen much faster.

If blood sugar levels are too high, then the hormone insulin is released. If levels drop below normal, the pancreas releases another hormone called glucagon.

This hormone will cause the glycogen that is stored in the liver to be converted back to glucose.

? Test Yourself

1. What is a hormone?
2. Where in the body is insulin made and released?
3. What effect does insulin have on the liver?
4. Write down one way that a person can be tested for diabetes.

★ Stretch Yourself

1. What is the difference in function between glucose, glucagon and glycogen?
2. Suggest what effect exercise is likely to have on the blood sugar level.

Organisms in Action

Homeostasis 2

Temperature Regulation

It is important to keep our body temperature at about **37°C**.

If the body temperature gets too low, this is called **hypothermia** which can be fatal. If the blood temperature gets too high, it could lead to **heat stroke** and **dehydration**.

The body temperature is monitored by the brain and if it varies from 37°C various changes are brought about.

When we feel hot we need to lose heat faster as our core body temperature is in danger of rising.

We do this by:
• **Sweating** – as water evaporates from our skin, it absorbs heat energy. This cools the skin and the body loses heat.

• Sending more blood to the skin so that more heat is lost by radiation. This causes the skin to look red.

When we feel too cold we are in danger of losing heat too quickly and cooling down. This means we need to conserve our heat to maintain a constant 37°C.

We do this by:
• **Shivering**, which is a rapid contraction and relaxation of body muscles. This increases the rate of respiration and more energy is released as heat.
• Sending less blood to the skin so the blood is diverted to deeper within the body to conserve heat. This causes the skin to look pale.
• Sweating less.

Build Your Understanding

It is important to keep the body temperature at 37°C because it is the best temperature for enzymes in the body to work. Enzymes control the rate of chemical reactions in the body and so are important molecules.

Any change in body temperature is detected by the thermoregulatory centre in the brain. This will bring about these correction mechanisms:

✓ Maximise Your Marks

It is important to describe vasodilation and vasoconstriction as the widening or narrowing of blood vessels. Lots of students lose marks because they say that blood vessels move towards or away from the skin.

Body temperature is detected in the thermoregulatory centre in the brain → If the temperature rises, various changes occur → Increased sweat production which evaporates and widened blood vessels in the skin, allowing blood to flow nearer the skin surface (**vasodilation**) → Body temperature returns to normal

If the temperature drops, changes occur to slow down heat loss → Decreased sweat production and closing down of blood vessels close to the skin (**vasoconstriction**) → Body temperature returns to normal

Control of Water Balance

It is important to control the amount of water in the body, otherwise the blood can become too concentrated or diluted.

This is done by making sure that over a certain period of time we take in the same amount of water as we give out.

Most of the regulation of water content is done by the **kidneys**, which alter the volume and concentration of the urine.

The body gains water by:
- drinking
- eating food
- respiring, which releases water.

The body loses water by:
- sweating
- breathing
- defecating
- excreting urine.

Build Your Understanding

The kidneys control the water balance of the body.

The kidneys do this by filtering the blood to remove all small molecules.

Then useful molecules (e.g. glucose), and a certain amount of water and salts, are taken back into the blood to keep their levels in balance.

The remaining waste is stored in the bladder as urine.

The amount of water that is taken back into the blood is controlled by a hormone called antidiuretic hormone (ADH) which is released by the pituitary gland.

Different drugs can alter ADH release:
- Alcohol reduces ADH release and can cause too much urine to be made.
- Ecstasy can have the opposite effect.

Warm temperatures, exercise, salt intake or lack of fluids	→	The blood becomes too concentrated	→	The pituitary gland releases more ADH	→	More water is re-absorbed and a more concentrated urine is made

❓ Test Yourself

1. What is the normal human body temperature?
2. Why do we look red when we are hot?
3. Why can becoming too hot lead to dehydration?
4. What is the main organ that controls water balance in the body?

⭐ Stretch Yourself

1. Frostbite is caused by being in very cold conditions for a long time. The cells in the fingers and toes die. Suggest why this happens.
2. Some people who have an inactive pituitary gland produce larger than normal quantities of urine. Explain why this happens.

Hormones and Reproduction

Organisms in Action

Reproductive Hormones

Hormones are responsible for controlling many parts of the reproduction process.

This includes controlling:
- The development of the sex organs.
- The production of sex cells.
- Pregnancy and birth.

The main hormones controlling these processes are shown in the table below.

Testosterone and **oestrogen** control the changes occurring in the male and female bodies at puberty. These changes are the **secondary sexual characteristics**.

Testosterone Body hair grows, voice breaks, muscle growth increases

Oestrogen Breasts grow, pubic hair grows, wide hips develop

Hormone	Male or female	Produced by	Main function
Testosterone	Male	Testes	Stimulates the male secondary sexual characteristics
Oestrogen	Female	Ovaries	Stimulates the female secondary sexual characteristics; repairs the wall of the uterus
Progesterone	Female	Ovaries and placenta	Prevents the wall of the uterus breaking down

Hormones and the Menstrual Cycle

Reproductive hormones also control the production of the sex cells. In males they are sperm and in females they are eggs.

After puberty in the male, sperm are produced continuously, but in the female one egg is usually released about once a month.

This means that oestrogen and **progesterone** levels vary at different times in the monthly or **menstrual cycle**:
- Oestrogen levels are high in the first half of the cycle. The oestrogen prepares the wall of the uterus to receive a fertilised egg. It does this by making it thicker and increasing its supply of blood before the egg is released (**ovulation**).
- Progesterone is high in the second half of the cycle. It further repairs the wall of the uterus and stops it breaking down.

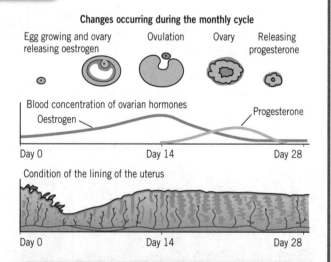

Changes occurring during the monthly cycle

Egg growing and ovary releasing oestrogen   Ovulation   Ovary   Releasing progesterone

Blood concentration of ovarian hormones
Oestrogen   Progesterone

Day 0   Day 14   Day 28

Condition of the lining of the uterus

Day 0   Day 14   Day 28

✔ Maximise Your Marks

If a question gives a diagram of the menstrual cycle and asks you when ovulation occurs, do not automatically say day 14. Ovulation occurs slightly after the oestrogen peak and before the progesterone one, when there is a spike in luteinising hormone. All women are different. In the diagram above this would be day 15.

Build Your Understanding

The production of oestrogen and progesterone is controlled by the release of other hormones. These hormones are called luteinising hormone (LH) and follicle stimulating hormone (FSH). Both are made in the pituitary gland in the brain.

Control of reproduction

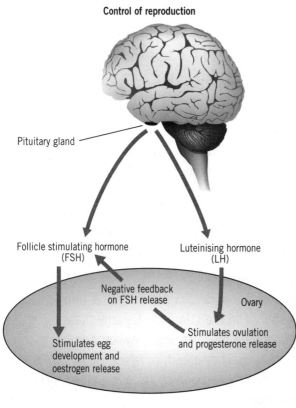

Pituitary gland

Follicle stimulating hormone (FSH)

Luteinising hormone (LH)

Negative feedback on FSH release

Ovary

Stimulates egg development and oestrogen release

Stimulates ovulation and progesterone release

💡 Boost Your Memory

There are five different hormones to remember that are involved in controlling reproduction. Copy out the table in the section on reproductive hormones (page 10) and see if you can add LH and FSH to it.

Treating Infertility

About one in seven of all couples have difficulty having a baby. There are many reasons for **infertility** and these can include:

- A blockage in the fallopian tubes or in the sperm ducts.
- The eggs not being developed or released from the ovaries.
- Not enough fertile sperm being produced by the testes.

It is now possible to treat some of these cases of infertility by using hormones:

- Women who do not develop or release eggs from their ovaries can take a **fertility drug**. This contains hormones that are similar to FSH. The drugs stimulate the production of eggs and sometimes a number of eggs are released each month.
- Women who have blocked fallopian tubes can be treated with fertility drugs and a number of eggs are removed from their body. These eggs can be fertilised by sperm outside the body. The embryo can then be put back inside the uterus. This process is called **in vitro fertilisation (IVF)**.

Decreasing Fertility

Some women may want to stop themselves becoming pregnant. They take drugs that are called **oral contraceptives**. These drugs contain different amounts of the hormones oestrogen and progesterone. They prevent the pituitary gland releasing FSH. This means that the ovary will not produce eggs.

❓ Test Yourself

1 Where is testosterone made and what does it do?

2 What is ovulation?

3 What is IVF?

4 How do oral contraceptives work?

⭐ Stretch Yourself

1 Normally the progesterone level falls towards the end of the menstrual cycle. Why is it important that it stays high if the egg has been fertilised?

2 The 'morning after pill' contains very high levels of oestrogen. Suggest why many people think that it should not be used regularly for contraception.

Responding to the Environment

Patterns of Response

All living organisms need to respond to changes in the environment. Although this happens in different ways, the pattern of events is always the same:

stimulus → detection → co-ordination → response

There are three main steps in this process:

Step 1: Detecting the stimulus
Receptors are specialised cells that detect a stimulus. Their job is to convert the stimulus into electrical signals in nerve cells. Some receptors can detect several different stimuli, but they are usually specialised to detect one type of stimulus.

Stimulus	Type of receptor
Light	Photoreceptors in the eyes
Sound	Vibration receptors in the ears
Touch, pressure, pain and temperature	Different receptors in the skin
Taste and smell	Chemical receptors in the tongue and nose
Position of the body	Receptors in the ears

A **sense organ** is a group of receptors gathered together with some other structures.

The other structures help the receptors to work more efficiently. An example of this is the eye.

Step 2: Co-ordination
The body is receiving information from many different receptors at the same time.

Co-ordination involves processing all the information from receptors so that the body can produce a response that will benefit the whole organism. In most animals this job is done by the **central nervous system (CNS)**, which is made up of the brain and spinal cord.

Step 3: Response
Effectors are organs in the body that bring about a response to the stimulus. Usually these effectors are muscles and they respond by contracting. They could, however, be glands and they may respond by releasing an enzyme or hormones.

An Example of a Sense Organ – The Eye

The light enters the eye through the **pupil**.

It is focused on to the **retina** by the **cornea** and the **lens**.

The size of the pupil can be changed by the muscles of the iris when the brightness of the light changes. The aim is to make sure that the same amount of light enters the eye all the time.

The job of the lens is to change shape so that the image is always focused on the light-sensitive retina.

The receptors are cells in the retina called **rods** and **cones**. They detect light and send messages to the brain along the **optic nerve**.

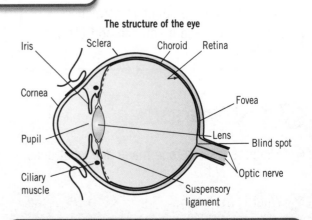

The structure of the eye

Iris, Sclera, Choroid, Retina, Cornea, Fovea, Pupil, Lens, Blind spot, Ciliary muscle, Suspensory ligament, Optic nerve

💡 Boost Your Memory

Remember that both the cornea and the lens refract light, but it is the job of the lens to make the fine adjustment to focus the light on the retina.

Build Your Understanding

The lens must be a different shape when the eye looks at a close object compared with when it looks at a distant object. This is to make sure that the light is always focused on the back of the retina.

The ciliary muscle changes the shape of the lens, as shown in the diagram. This is called accommodation.

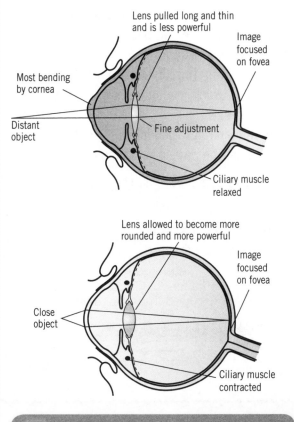

Lens pulled long and thin and is less powerful

Image focused on fovea

Most bending by cornea

Distant object

Fine adjustment

Ciliary muscle relaxed

Lens allowed to become more rounded and more powerful

Image focused on fovea

Close object

Ciliary muscle contracted

Eye Defects

Some people have problems with their eyes. There are a number of different causes:

Condition	Cause	Treatment
Long-sight or short-sight	The eyeball or lens is the wrong shape	Long-sight can be corrected by wearing convex lenses; short-sight can be corrected by wearing concave lenses; cornea surgery can also be used
Red-green colour blindness	Lack of certain cones in the retina	No treatment
Poor accom.	Lens becomes less elastic in older people	Wearing glasses with half convex and half concave lenses

Judging Distance

The eyes are also used to judge distances:
- Animals that hunt usually have their eyes at the front of their head. Each eye has a slightly different image of the object. This is called binocular vision and this can be used to judge distance.
- Animals that are hunted usually have eyes on the side of their heads. This gives monocular vision and they cannot judge distances so well. They can, however, see almost all around.

Organisms in Action

The Nervous System

Organisms in Action

Nerves and Neurones

To communicate between receptors and effectors the body uses two main methods. These are:
- neurones
- nerves.

A **neurone** is a single specialised cell that is adapted to pass electrical impulses.

The brain and spinal cord contain millions of neurones. Outside the CNS, neurones are grouped together into bundles of hundreds or thousands. These bundles are called **nerves**.

The three main types of neurones are:
- **Sensory neurones**, which carry impulses from the receptors to the CNS.
- **Motor neurones**, which carry impulses from the CNS to the effectors.

- **Relay neurones**, which pass messages between neurones in the CNS.

Although all neurones have different shapes, they all have certain features in common:
- One or more long projections (**axons** and **dendrons**) from the cell body to carry the impulse a long distance.
- A fatty covering (**myelin sheath**) around the projection, which insulates it and speeds up the impulse.
- Many fine endings (**dendrites**) so that the impulse can be passed on to many cells.

Each neurone does not directly end on another neurone.

There is a small gap between the two neurones and this is called a **synapse**.

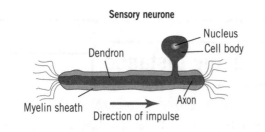

Sensory neurone

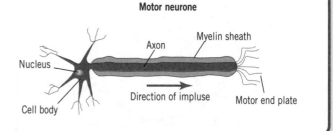

Motor neurone

Build Your Understanding

So that an impulse can be generated in the next neurone, a chemical called a neurotransmitter is released when the nerve impulses reaches the synapse.

This then diffuses across the small gap and joins with receptors on the next neurone. This starts a nerve impulse in this cell.

Many drugs work by interfering with synapses. They may block or copy the action of the neurotransmitters in certain neurones.

Some of these are discussed on page 26.

Chemical transmission between nerves

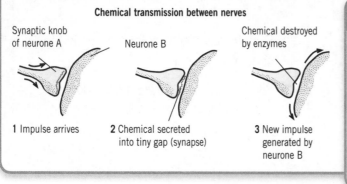

1 Impulse arrives

2 Chemical secreted into tiny gap (synapse)

3 New impulse generated by neurone B

✔ Maximise Your Marks

To get an A* you must be able to explain why certain molecules might affect the working of synapses. Neurotransmitters have a specific shape to fit into receptors, so other molecules that have a similar shape might block the site or even act like the transmitter.

Reflex Responses

The **peripheral nervous system** is made up of all the nerves that pass information to and from the CNS. Once impulses reach the CNS from a sensory neurone there is a choice:

- Either the message is passed straight to a motor neurone via a relay neurone. This is very quick and is called a **reflex action**.
- Or the message is sent to the higher centres of the brain and the organism might decide to make a response. This is called a **voluntary action**.

All reflexes:
- are fast
- do not need conscious thought
- protect the body.

Examples of reflexes include the knee jerk, pupil reflex, accommodation, ducking, and withdrawing the hand from a hot object.

This diagram shows the pathway for a reflex that involves the spinal cord. The numbers in the diagram refer to the flow diagram opposite.

A reflex action

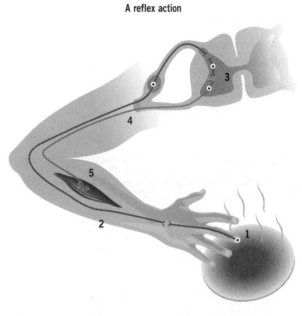

1 Stimulus is detected by sensory cell.

⬇

2 Impulse passes down sensory neurone.

⬇

3 Relay neurone passes impulse to motor neurone.

⬇

4 Motor neurone passes impulse to effector (muscle).

⬇

5 Muscle contracts.

Organisms in Action

💡 Boost Your Memory

Use this short mnemonic to remember the order of events in a reflex action: **S**ue **r**emembers **s**eeing **R**achel make **e**gg **r**olls.

This stands for: **s**timulus, **r**eceptor, **s**ensory neurone, **r**elay neurone, **e**ffector, **r**esponse

✔ Maximise Your Marks

Students often lose marks because they say that reflexes 'do not involve the brain'. Some do not, but some such as blinking do. None of them involves conscious thought.

❓ Test Yourself

1. What is the difference between a nerve and a neurone?
2. What is the job of the fatty sheath around a neurone?
3. What is a synapse?
4. Why is it important to the body that reflexes are fast?

⭐ Stretch Yourself

1. Scientists may want to produce a drug that is a painkiller. How can they use their knowledge of neurones and synapses to do this?
2. It is important that the body breaks down neurotransmitter molecules once they have stimulated a nerve impulse in the next neurone. Why is this?

Plant Responses

Plant Responses

Plants can also respond to changes in the external environment. These responses are usually slower than animal responses and include:

- Roots and shoots growing towards or away from a particular stimulus.
- Plants flowering at a particular time.
- The ripening of fruits.

The type of response that involves part of the plant growing in a particular direction is called a **tropism**.

If the growth is in response to gravity, it is a **geotropism (gravitropism)**.

If it is in response to light, it is a **phototropism**.

Stimulus	Growth of shoots	Growth of roots
Gravity	Away from = negatively geotropic	Towards = positively geotropic
Light	Towards = positively phototropic	Away from = negatively phototropic

By controlling the growth of plants, auxins (plant hormones) can allow plants to respond to changes happening around them. This means that the roots and shoots of plants can respond to gravity or light in different ways.

These responses help the shoot to find light for photosynthesis and the root to grow down to anchor the plant in the soil and absorb water and mineral ions.

Growth is in the direction of the light source

✓ Maximise Your Marks

Remember that it is important for shoots to grow away from gravity.

When a seed germinates, it is often under the soil, so the shoot cannot grow towards light as it is dark.

Build Your Understanding

Auxins are responsible for making plant cells increase in length or elongate.

Experiments have shown that a series of steps are involved in the response:

- Light is shone on one side of a shoot.
- More auxin is sent down the side of the shoot that is in the shade.

- This causes cells on the shaded side to elongate more.
- The shoot therefore grows towards the light.

How auxins control tropisms

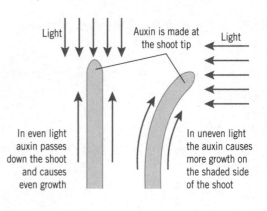

Light | Auxin is made at the shoot tip | Light

In even light auxin passes down the shoot and causes even growth

In uneven light the auxin causes more growth on the shaded side of the shoot

✓ Maximise Your Marks

To get an A* you must be able to look at experiments on shoots and work out which way they will grow. Learn and practise this skill by trying question 4 on page 19.

Plant Hormones

Growth in plants is controlled by chemicals called plant growth substances or **plant hormones**. There are a number of different types, but the main types are called **auxins**.

Auxins are made in the tip of the shoot and move through the plant acting as the signal to make the shoot grow towards the light.

Auxins change the direction that roots and shoots grow by changing the rate that the cells elongate.

Gardeners can use plant hormones such as auxins to help them control the growth of their plants.

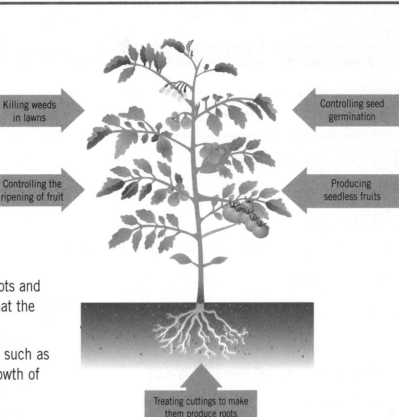

Killing weeds in lawns

Controlling seed germination

Controlling the ripening of fruit

Producing seedless fruits

Treating cuttings to make them produce roots

Build Your Understanding

The plant hormones used by gardeners work in a number of ways:

- Auxins kill weeds by making them grow too fast so that they use up all their food reserves. These weedkillers can be selective because they only kill broad leaf weeds, not grass.
- Some seeds need to be in the soil for some time before they germinate. Plant hormones can speed up this process.
- Fruit can be picked unripe so that it can be transported without being damaged. Then it can be ripened using plant hormones.

? Test Yourself

1. Shoots are positively phototropic. What does this mean?

2. Roots are positively phototropic. What does this mean and why is it important for a plant?

3. Where are auxins made in a plant?

4. Why are shoots dipped into hormone powder when taking cuttings?

★ Stretch Yourself

1. Placing a foil cap over the tip of a shoot allows it to grow but stops it bending towards light. Explain why.

2. Write down one advantage and one disadvantage of producing seedless fruits.

Practice Questions

 Complete these exam-style questions to test your understanding. Check your answers on page 120. You may wish to answer these questions on a separate piece of paper.

1 Complete this passage about a balanced diet by writing the correct words in the gaps. (4)

A balanced diet contains _____ main groups of food chemicals. One group is protein,

which is necessary for _____. Proteins are made from molecules called

_____. Minerals are another group of chemicals. An example is _____,

which is needed to make haemoglobin.

2 A new device is being developed by scientists to treat diabetics.

A = Pump delivers a dose of insulin.

B = Insulin is injected through thin tube.

C = Blood glucose level is constantly measured.

D = Transmitter sends information about blood glucose level to pump.

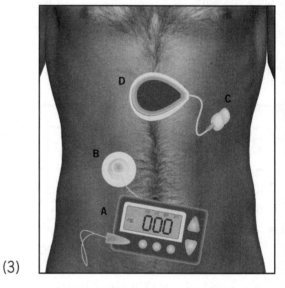

Diabetics currently have to take blood samples to test for how much insulin to inject.

a) Explain how insulin helps to regulate the glucose levels in the blood. (3)

....

....

....

b) Why is it necessary for diabetics to alter their dose of insulin at different times? (1)

....

....

....

c) Scientists think that the new system could be an improvement on the current method of testing blood samples. Suggest why. (1)

....

....

....

....

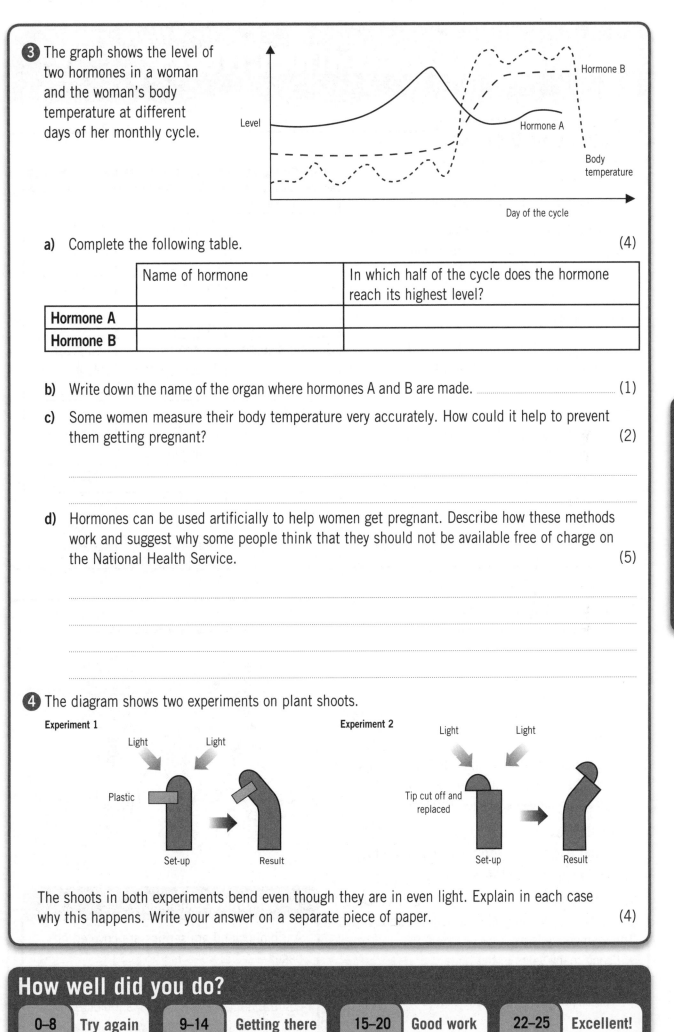

3 The graph shows the level of two hormones in a woman and the woman's body temperature at different days of her monthly cycle.

Level

Hormone B

Hormone A

Body temperature

Day of the cycle

a) Complete the following table. (4)

	Name of hormone	In which half of the cycle does the hormone reach its highest level?
Hormone A		
Hormone B		

b) Write down the name of the organ where hormones A and B are made. .... (1)

c) Some women measure their body temperature very accurately. How could it help to prevent them getting pregnant? (2)

....

....

d) Hormones can be used artificially to help women get pregnant. Describe how these methods work and suggest why some people think that they should not be available free of charge on the National Health Service. (5)

....

....

....

....

4 The diagram shows two experiments on plant shoots.

Experiment 1

Light   Light

Plastic

Set-up   Result

Experiment 2

Light   Light

Tip cut off and replaced

Set-up   Result

The shoots in both experiments bend even though they are in even light. Explain in each case why this happens. Write your answer on a separate piece of paper. (4)

How well did you do?

| 0–8 | Try again | 9–14 | Getting there | 15–20 | Good work | 22–25 | Excellent! |

Pathogens and Infections

Causes of Disease

A disease occurs when the normal functioning of the body is disturbed.

Infectious diseases can be passed on from one person to another, but **non-infectious diseases** cannot.

Organisms that cause infectious diseases are called **pathogens**.

There are a number of different types of organism that can be pathogens, as shown in the table.

Pathogens may reproduce rapidly in the body by either damaging cells directly or producing chemicals called toxins which make people feel ill.

Viruses damage cells by taking over the cells and reproducing inside them. They then burst out of the cells destroying them in the process.

Type of disease	Description	Examples
Non-infectious		
Body disorder	Incorrect functioning of a particular organ	Diabetes, cancer
Deficiency disease	Lack of a mineral or vitamin	Anaemia, scurvy
Genetic disease	Caused by a defective gene	Red-green colour blindness
Infectious		
	Type of pathogen:	Examples
	Fungi	Athlete's foot
	Viruses	Flu
	Bacteria	Cholera
	Protozoa	Malaria

Spreading Disease

If the pathogens do enter the body then the body will attack them in a number of ways.

The area that is infected will often become inflamed and two types of white blood cells attack the pathogen.

Pathogens are detected by the white blood cells because the pathogens have foreign chemical groups called **antigens** on their surface.

The **antibodies** that are produced are specific to a particular pathogen or toxin and will only attach to that particular antigen.

When an antigen is detected by white blood cells, they produce memory cells as well as antibodies. The memory cells work by:
- Living many years in the body.
- Producing antibodies very quickly if the same type of pathogen reinvades the body.

The actions of white blood cells

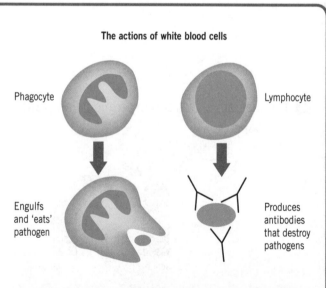

Phagocyte

Lymphocyte

Engulfs and 'eats' pathogen

Produces antibodies that destroy pathogens

✔ Maximise Your Marks

Antibiotics and antiseptics are covered on pages 22–23.

Students often lose marks because they confuse antibodies, antigens, antibiotics and antiseptics. Make sure you know the difference between them all.

The Entry of Pathogens

There are a number of different ways that pathogens can be spread from one person to another.

The skin covers most of the body and is quite good at stopping pathogens entering the body.

The body has a number of other defences that it uses in order to try to stop pathogens entering.

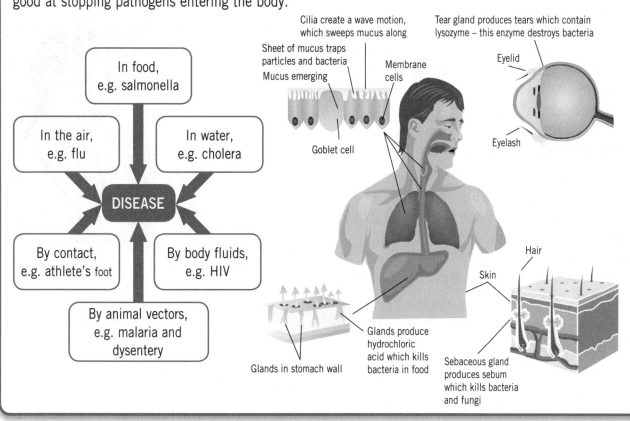

The body's defences

In food, e.g. salmonella

In the air, e.g. flu

In water, e.g. cholera

DISEASE

By contact, e.g. athlete's foot

By body fluids, e.g. HIV

By animal vectors, e.g. malaria and dysentery

Cilia create a wave motion, which sweeps mucus along

Sheet of mucus traps particles and bacteria

Mucus emerging

Membrane cells

Goblet cell

Tear gland produces tears which contain lysozyme – this enzyme destroys bacteria

Eyelid

Eyelash

Hair

Skin

Glands in stomach wall

Glands produce hydrochloric acid which kills bacteria in food

Sebaceous gland produces sebum which kills bacteria and fungi

Build Your Understanding

By studying how pathogens are spread from person to person it is possible to find ways of preventing this spread.

This will help to reduce the number of people getting the disease.

For example, the malaria pathogen is spread by mosquito bites. Therefore using insect repellents, insect nets and draining swamps can reduce cases of malaria.

Make sure that you know which diseases are mentioned on your specification, which type of organism causes each disease, and how they are spread from person to person.

? Test Yourself

1. Write down one example of a deficiency disease.
2. What type of organism causes malaria?
3. What is an antibody?
4. What is a phagocyte?

★ Stretch Yourself

1. Releasing thousands of sterile male mosquitoes into the environment helps reduce the incidence of malaria. Explain why.
2. Why does a person usually get measles only once?

Antibiotics and Antiseptics

Antibiotics

Sometimes a pathogen can produce illness before our body's immune system can destroy it. It is sometimes possible for us to take drugs called **antibiotics** to kill the pathogen:

- Antibiotics are chemicals that are usually produced by microorganisms, especially fungi
- Antibiotics kill bacteria and fungi but do not have any effect on viruses.

The first antibiotic to be widely used was penicillin, but there are now a number of different antibiotics that are used to treat different bacteria. This has meant that some diseases that once killed millions of people can now be treated.

There is a problem, however. More and more strains of bacteria are appearing that are resistant to antibiotics.

There are various ways that doctors try to prevent the spread of these resistant bacteria:

"I tell my patients to finish the dose of antibiotics even if they feel better."

"I regularly change the antibiotics that I prescribe and sometimes use combinations of different antibiotics."

"I prescribe antibiotics only in serious cases caused by bacteria."

"I always wash my hands with antiseptic between patients."

Build Your Understanding

Antibiotic resistance first appears due to a genetic change or mutation, and soon afterwards a large population of resistant bacteria can appear.

This process has occurred in many different types of bacteria including the TB-causing bacterium and one called MRSA. These bacteria are now resistant to many different types of antibiotic and so are very difficult to treat.

The development of antibiotic resistance

Antibiotics added to bacteria

Antibiotics kill off all susceptible bacteria

Only a single resistant bacterium survives

Bacterium multiplies to produce millions of resistant bacteria

New resistant super-bugs

Antiseptics

One important weapon against resistant bacteria is the use of **antiseptics**:

- Antiseptics are man-made chemicals that kill pathogens outside the body.
- They were first used by an Austrian doctor called Dr Semmelweis to sterilise medical instruments
- They are widely used in hospitals to try to prevent the spread of resistant bacteria.

An antiseptic is usually used on the body, and a disinfectant is usually used on other surfaces.

⚡ Boost Your Memory

Draw a spider diagram in your revision book to show antibiotics, antiseptics, antibodies and disinfectants. Make sure it shows what they kill and what makes them.

Testing Antibiotics and Antiseptics

It is possible to grow microorganisms such as bacteria in laboratories.

They are grown on a special jelly called **agar** in a Petri dish.

The agar is a culture medium containing an energy source, minerals and sometimes vitamins and protein.

Certain precautions have to be taken (see diagram below):

It is then possible to see what action certain antibiotics or antiseptics have on the microorganisms.

Filter paper discs can be soaked in different antibiotics and placed on the agar.

The more effective the antibiotic, the wider the area of bacteria that will be killed.

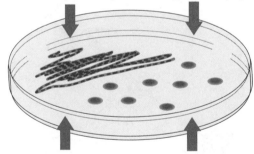

Growing bacteria

Petri dishes must be sterilised before use

The lid of the Petri dish must be sealed with tape to prevent microorganisms escaping or entering

Inoculating loops used to transfer microorganisms must be sterilised in a flame

In school laboratories Petri dishes must be incubated at a maximum of 25°C to reduce the growth of harmful microorganisms

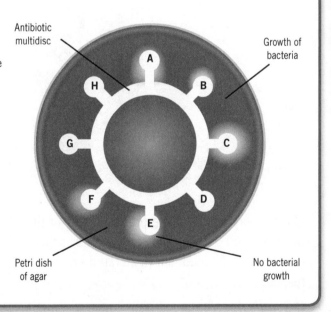

Antibiotic multidisc

Growth of bacteria

Petri dish of agar

No bacterial growth

❓ Test Yourself

1. Why cannot antibiotics be used to prevent flu?
2. Write down the name of one antibiotic.
3. Write down one difference between an antibiotic and an antiseptic.
4. Why is it important to incubate Petri dishes at a maximum of 25°C in school laboratories?

★ Stretch Yourself

1. Suggest why microorganisms such as fungi might produce antibiotics.
2. Explain why doctors may prescribe a combination of different antibiotics to treat a patient.

Vaccinations

What is a Vaccine?

When our body encounters a pathogen, white blood cells make antibodies against the pathogen.

If they encounter the same pathogen again in the future then antibodies are produced faster and the pathogen is killed more quickly. This is called **immunity**.

This idea has been used in **vaccinations**.

A vaccine contains harmless versions of the pathogen which stimulate immunity.

Build Your Understanding

The harmless version of the pathogen contained in the vaccine could be:

- Dead pathogens.
- Live but weakened pathogens.
- Parts of the pathogen that contain antigens.

These all contain the specific antigens that are detected by the body's white blood cells.

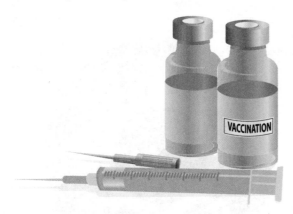

The memory cells that are produced stay in the body and will detect identical antigens in the future. This will lead to a more rapid immune response.

If a new strain of the pathogen appears then the current vaccination may not be effective.

✓ Maximise Your Marks

Questions often ask why it is necessary to produce a different flu vaccine every year. It is because different strains of the flu virus appear at regular intervals and they have different antigens. This means that the current memory cells would not recognise them.

Active and Passive Immunity

The type of immunity, in which the antibodies are made by the person is called **active immunity**.

Sometimes it might be too late to give somebody this type of vaccination because they already have the pathogen.

They can be given an injection containing antibodies made by another person or animal. This is called **passive immunity**. It gives quicker protection but it does not last as long.

A vaccination containing antibodies

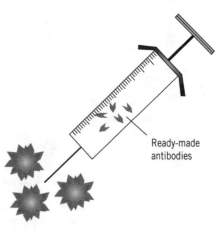

Ready-made antibodies

Passive immunity also occurs when a baby receives antibodies from its mother across the placenta or in breast milk.

💡 Boost Your Memory

Make a table to show the differences between active and passive immunity. Include who produces the antibodies, an example of when each might occur, how quickly they work, and how long they last for.

The Risks of Vaccines

Whether or not to give children vaccinations is a difficult decision for some people to make. Diseases like measles, mumps and rubella can have serious effects on the body:

- Measles is a very serious disease – 1 in 2500 babies that catch the disease die.
- Mumps may cause deafness in young children.
- Mumps may also cause viral meningitis which can be fatal.
- Rubella can cause a baby to have brain damage if its mother catches the disease during pregnancy.

The introduction of a combined **measles, mumps and rubella (MMR)** vaccine has led to a decrease in measles.

In 1998 a study of autistic children raised the question of a connection between the MMR vaccine and autism.

This led to a decrease in the number of parents allowing their children to have the MMR vaccine.

This study has been discredited, but it is impossible to say that having a vaccine does not involve a **risk**.

Some people say that parents should be forced to allow their children to have the vaccine, otherwise the disease will not disappear. Others say that it should be a personal choice.

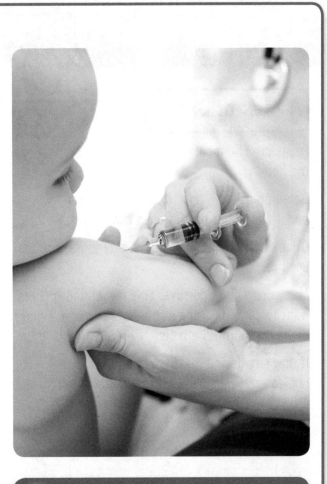

✓ Maximise Your Marks

Exam papers often ask questions about the risks of vaccines as this is part of *How Science Works*. Make sure you realise that taking any drug carries a risk and that people will react differently. However, like many things in life, it is a question of balancing risks, in this case the risks of the vaccination against the risk of the disease.

❓ Test Yourself

1. What does a vaccine contain?
2. What type of immunity is produced in the following cases?
   a) Antibodies are taken from a horse that has rabies and injected into a person.
   b) A person has chicken-pox and is now immune to this disease.
3. Explain your answers to question 2.

⭐ Stretch Yourself

1. Why do people sometimes feel ill after having certain vaccines?
2. Some pathogens can change their antigens during their life. What effect does this have?

Drugs

Types of Drugs

Drugs are chemicals that alter the functioning of the body.

Some drugs such as antibiotics are often beneficial to our body if used correctly.

Others can be harmful, particularly those that are used recreationally.

Many drugs are **addictive**. This means that people want to carry on using a drug even though it may be having harmful effects.

If they stop taking the drug, they may suffer unpleasant side effects called **withdrawal symptoms**.

It also means that people develop **tolerance** to the drug, which means that they need to take bigger doses for it to have the same effect. Heroin and cocaine are very addictive.

In order to control drugs, many can only be bought with a prescription.

Drugs are also classified into groups. Class A drugs are the most dangerous, with class C being the least dangerous. If people are caught with illegal class A drugs, the penalties are the highest.

Different drugs do different things

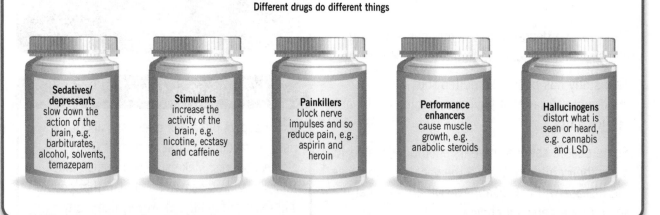

Sedatives/ depressants slow down the action of the brain, e.g. barbiturates, alcohol, solvents, temazepam

Stimulants increase the activity of the brain, e.g. nicotine, ecstasy and caffeine

Painkillers block nerve impulses and so reduce pain, e.g. aspirin and heroin

Performance enhancers cause muscle growth, e.g. anabolic steroids

Hallucinogens distort what is seen or heard, e.g. cannabis and LSD

Testing New Drug Treatments

Any new drugs have to be tested before they are used on patients. Doctors need to know:
- if the treatment works
- if it is safe

There are a number of different ways that a new treatment can be tested:
- First it is tested on cells in a laboratory. This is on human cells to see if it is harmful and on microorganisms in Petri dishes to see if it will kill them.
- If it passes these tests the drug is then tried on animals.
- Then the drug is tested on healthy human volunteers for safety, and on people with the illness for safety and effectiveness.

Many of these tests cause disagreements. Many people think that animals should not be used to test drugs. Some think that it is too cruel, while others think that it is pointless as the effects may be different on animals from on humans.

Others think that the tests are reasonable because the benefits outweigh the risks of the tests.

Boost Your Memory

Exam questions often ask about testing drugs on animals or volunteers as this is an ideal *How Science Works* subject. Be prepared to give both sides of the argument, even if you feel strongly one way or the other.

Build Your Understanding

Stimulants such as nicotine affect synapses (the junction of two neurones) by causing more neurotransmitter substances to cross to the next neurone and bind to the receptor molecules.

This makes it more likely for an impulse to be conducted in the next neurone.

Depressants such as alcohol bind with receptor molecules, thereby blocking the transmission of impulses.

Some Controversial Drugs

Over the years the use of some drugs has been particularly controversial.

Thalidomide is a drug that was given to pregnant women to try to relieve morning sickness. Thalidomide had been tested on pregnant animals.

Unfortunately, many babies born to mothers who took the drug were born with severe limb abnormalities. The drug was then banned.

More recently, thalidomide has been used successfully in the treatment of leprosy and other diseases.

Cannabis is an illegal drug. Many people have argued about whether it should be a class B or class C drug or possibly made legal. Cannabis smoke does contain harmful chemicals which may cause mental illness in some people.

Double Blind Tests

Once a drug is cleared to be tested on patients, the trial has to be set up carefully.

One group is given the drug and another group has a **placebo**. A placebo looks like the real treatment but has no drug in it.

If the two groups do not know which treatment they are having but the doctor does, then this is called a **blind test**. In a blind test the researcher may give clues to the volunteers without realising it. This can make the results unreliable.

If, in addition, the doctor does not know then this is called a **double blind test**. It means that the people involved are not influenced by knowing which treatment is being given.

This removes the opportunity for bias and the tests can be more reliable. Double blind tests are more complex to set up though.

Some people think that placebos should not be used in tests on ill people. They say that it is not right to make people believe that they are receiving a possible cure when they are not.

Others think that the tests are reasonable because the benefits outweigh the risks of the tests.

✔ Maximise Your Marks

You must be able to analyse how a drugs test is done and decide if it is an open test, a blind test or a double blind test. Look at who knows which the real drug is.

❓ Test Yourself

1 Why is it difficult to stop taking drugs such as cocaine?

2 Why are illegal drugs put into different classes?

3 Why are drugs tested on healthy volunteers?

4 Why do some people object to using animals to test drugs?

⭐ Stretch Yourself

1 Before anaesthetics were available, surgeons often gave patients brandy to drink before operations. Suggest why they did this.

2 Explain why it is important that the doctor treating patients does not know whether the patients are taking the real drug or a placebo.

Smoking and Drinking

Smoking

Many people cannot give up smoking tobacco because it contains the drug **nicotine**. This is addictive. The nicotine is harmful to the body, but most damage is done by the other chemicals in the tobacco smoke.

As well as the effects described in the diagram, smoking can increase blood pressure. It does this in two main ways:

- Nicotine increases the heart rate directly.
- Carbon monoxide reduces the oxygen-carrying capacity of the blood by combining with haemoglobin. This causes the heart rate to increase to compensate.

Problems resulting from smoking

- The heat and chemicals in the smoke destroy the cilia on the cells lining the airways. The goblet cells also produce more mucus than normal. The bronchioles may become infected. This is called **bronchitis**.

- Chemicals in the tar may cause cells in the lung to divide uncontrollably. This can cause **lung cancer**.

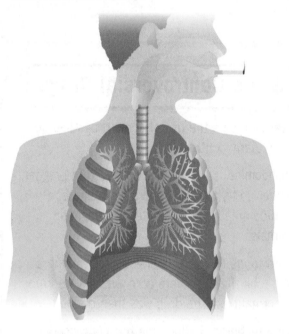

- The mucus collects in the alveoli and may become infected. This may lead to the walls of the alveoli being damaged. This reduces gaseous exchange and is called **emphysema**.

- The nicotine can cause an increase in blood pressure increasing the chance of a **heart attack**.

Smoking and Pregnancy

Smoking tobacco is particularly dangerous for pregnant women.

Women who smoke when they are pregnant are more likely to give birth to babies that have a low birth mass.

✓ Maximise Your Marks

The graph shows a good example of a **correlation**. Although there is quite a spread in the data, the trend shows that the more cigarettes a woman smokes, the lighter her baby is likely to be. You may be given graphs showing similar trends with smoking and lung cancer or heart disease.

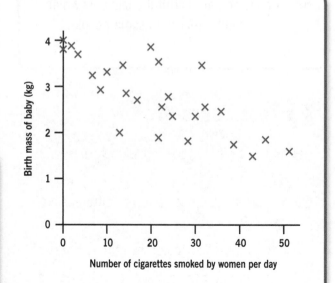

Number of cigarettes smoked by women per day

Drinking Alcohol

Drinking alcohol can have a number of effects on the body.

Short-term effects include:
- Upset balance and muscle control.
- Blurred vision and speech.

Long-term effects include:
- Damage to the liver (cirrhosis).
- Brain damage.
- Heart disease.

Different drinks have different concentrations of alcohol. To help people judge how much alcohol they have drunk, drinks are described as having a certain number of units of alcohol.

A single measure of spirits or a half pint of beer contains one unit of alcohol.

Due to the effects of alcohol on the body there is a legal limit for the level of alcohol in the blood of drivers and pilots.

Build Your Understanding

Drinking alcohol increases reaction times.

Alcohol slows down the bodies responses, so reaction times are increased, i.e. it takes you longer to respond.

✓ Maximise Your Marks

When answering questions about reaction times and stopping distances, you must remember that alcohol will increase reaction times and increase stopping distances. Candidates often get confused by this.

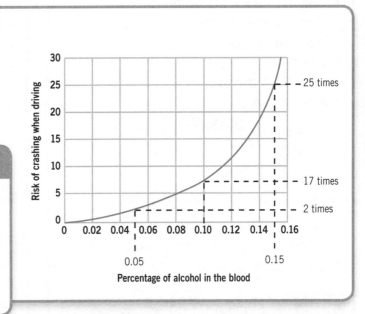

Percentage of alcohol in the blood

? Test Yourself

1. What effect does smoking have on mucus production in the lungs?

2. Why is gaseous exchange reduced in emphysema?

3. What name is given to the damage caused to the liver by alcohol?

4. How many units of alcohol are there altogether in two pints of beer and two single whiskeys?

★ Stretch Yourself

1. If a person smokes, the oxygen content of the blood drops. Why is it necessary for their heart rate to increase to compensate?

2. Using the graph above how many times more likely is a person going to have an accident if their blood alcohol level is 0.13% compared with 0.08%?

Too Much or Too Little

Eating Too Much

It is important to maintain a balanced diet for the healthy functioning of the body.
In the developed world many people eat too much food.

This can make a person more likely to get various diseases.

If a person eats food faster than it is used up by the body, then the excess will be stored. Much of this will be stored as fat and can lead to **obesity**.

Obesity can be linked to a number of different health risks:
- Arthritis – the joints wearing out.
- Type 2 diabetes – inability to control the blood sugar level.
- Breast cancer.
- High blood pressure.
- Heart disease.

It is possible to estimate if a person is underweight, ideal weight, overweight or obese by using the formula:

$$\text{Body Mass Index (BMI)} = \frac{\text{mass in kg}}{(\text{height in metres})^2}$$

The BMI figure can then be checked in a table to see what weight range a person is in.

BMI	What it Means
<18.5	Underweight – too light for your height
18.5–25	Ideal – correct weight range for your height
25–30	Overweight – too heavy for your height
30–40	Obese – much too heavy, health risks!

Blood Pressure

Contractions of the heart pump blood out into the arteries under pressure. This is so that the blood can reach all parts of the body.

Doctors often measure the blood pressure in the arteries and give two figures, for example 120 over 80. The higher figure is called the **systolic pressure** and this is the pressure when the heart contracts. The lower figure is when the heart is relaxed and this is the **diastolic pressure**.

Blood pressure varies depending on various factors.

The following factors can increase blood pressure:
- high salt in the diet
- high fat in the diet
- stress
- lack of exercise
- obesity
- high alcohol intake
- ageing.

Build Your Understanding

If left untreated, high blood pressure can cause various problems:
- Small blood vessels may burst because of the high pressure. If a small blood vessel bursts in the brain, it is called a stroke. Brain damage from a stroke can result in some paralysis and loss of speech.
- If a small blood vessel bursts in a kidney, the kidney may be damaged.

Low blood pressure can cause problems such as:
- Poor circulation.
- Dizziness and fainting, because the blood will not be at a high enough pressure to carry enough food and oxygen to the brain.

Heart Disease

The heart is made up of muscle cells that need to contract throughout life.

This needs a steady supply of energy, so the cells need oxygen and glucose at all times for respiration. This is supplied by blood vessels.

Fatty deposits called **plaques** can form in the walls of these blood vessels and reduce the flow of oxygen and glucose to the heart muscle cells.

This causes heart disease and if an area of muscle stops beating then this is a **heart attack**.

There are many factors that can make it more likely for a person to have heart disease.

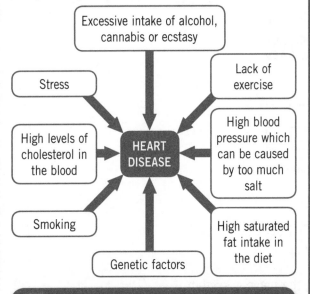

✓ Maximise Your Marks

In heart disease the blood vessels supplying the heart **muscle** are blocked.

Lots of students lose marks because they say that fat blocks up the blood vessels bringing back blood to the heart.

Eating Too Little

Eating too little of one type of food substance can lead to a deficiency disease.

Examples of deficiency diseases include:
- Anaemia due to a lack of iron.
- Scurvy due to a lack of vitamin C.
- Kwashiorkor due to a lack of protein.

Details of protein requirements are on page 4.

There are times when people do not eat enough food although there is food available. They may be on a diet because they have a poor self image or think that they are overweight when they are not. This can reduce their resistance to infection and cause irregular periods in women. It may lead to illnesses such as **anorexia**.

Coronary Thrombosis

Plaques in the walls of the **coronary arteries** supplying the heart muscle make it more likely that blood will start to clot in the blood vessels. A blood clot inside the vessel is called a **thrombosis**. This may block the blood vessel and cause a heart attack.

Drugs such as statins can be taken to reduce the levels of cholesterol in the blood.

❓ Test Yourself

1. A person has a mass of 80 kg and is 1.7 m tall. What is their BMI?
2. Why does a person's blood pressure have two figures?
3. What is a plaque?
4. Write down two ways that a person can try to reduce their risk of heart disease.

⭐ Stretch Yourself

1. People with low blood pressure may often experience cold fingers and toes. Suggest why.
2. People who are at risk of heart disease take drugs such as warfarin. This makes the blood less likely to clot. Why do they take this drug?

Practice Questions

Complete these exam-style questions to test your understanding. Check your answers on page 121. You may wish to answer these questions on a separate piece of paper.

Check your answers on page 121.

1 Fill in the gaps in the following sentences. Choose your words from this list: (4)

acid   antibodies   lysozyme   pathogen   sebum   toxin

An organism that causes a disease is called a _____1_____. The body tries to prevent these

organisms entering the body. Tears contain _____2_____ and the stomach makes

_____3_____, both of which can destroy the organism. If the organism enters the body,

it can damage cells or release poisons called _____4_____.

2 The following table gives some information about four drugs.

Drug	Type of action	Addictive
Anabolic steroids	Performance enhancer	No
Aspirin	Painkiller	No
Barbiturate	Depressant	Yes
Nicotine	Stimulant	Yes

a) Which drug shown in the table is found in cigarette smoke? (1)

b) Two of the drugs in the table are addictive. What does this mean? (2)

c) Suggest what action barbiturates may have on the nervous system. (2)

d) Barbiturates are a class B drugs and anabolic steroids are class C. Suggest why they are classified differently. (2)

3 The MMR vaccination is usually given to young children. It protects them from measles, mumps and rubella. Here is some information about these three diseases:

Measles is a very serious disease – 1 in 2500 babies that catch the disease die.

Mumps can cause deafness in young children.

Rubella can cause a baby to have brain damage if its mother catches the disease during pregnancy.

a) The MMR vaccination is given to children when they are very young rather than waiting until they are older. Write down one reason why. (1)

b) Pregnant women are tested to see if they have had the MMR vaccination. Why is this important? (1)

c) Explain how a vaccination like MMR can protect a person from getting a disease. (3)

....

....

4 a) Suggest two reasons why drugs are tested on animals. (2)

....

....

b) Explain why placebos are used when testing drugs. (2)

....

....

c) In the 1960s a drug called thalidomide was given to pregnant women. The drug had been tested on animals and seemed to be safe. Soon it became clear that the drug was causing the women to give birth to children with birth defects. Some animal rights campaigners say that this shows that drug testing on animals is a waste of time. Write down an argument for and against this view. (2)

....

....

5 The graph shows the number of people dying from heart disease in different countries and the amount of milk drunk in those countries.

a) Describe the trend shown by the graph. (1)

....

b) Suggest an explanation for this trend. (4)

....

....

How well did you do?

| 0–8 | Try again | 9–15 | Getting there | 16–22 | Good work | 23–27 | Excellent! |

Genes and Chromosomes

What is a Gene?

Most cells contain a nucleus that controls all of the chemical reactions that go on in the cell. Nuclei can do this because they contain the genetic material.

Genetic material controls the characteristics of an organism and is passed on from one generation to the next.

The genetic material is made up of structures called chromosomes. They are made up of a chemical called **deoxyribonucleic acid** or **DNA**.

The DNA controls the cell by coding for the making of proteins, such as enzymes. The enzymes will control all the chemical reactions taking place in the cell.

A **gene** is part of a chromosome that codes for one particular protein.

DNA codes for the proteins it makes by the order of four chemicals called **bases**. They are given the letters **A**, **C**, **G** and **T**.

By controlling cells, genes therefore control all the characteristics of an organism.

Different organisms have different numbers of genes and different numbers of chromosomes. In most organisms that reproduce by sexual reproduction, the chromosomes can be arranged in pairs. This is because one of each pair comes from each parent.

Chromosomes and Reproduction

No living organism can live forever so there is a need to reproduce. **Sexual reproduction** involves the passing on of genes from two parents to the offspring. This is why we often look a little like both of our parents.

The genes are passed on in the **sex cells** or **gametes** which join at **fertilisation**. In humans, each body cell has 46 chromosomes in 23 pairs. This means that when the male sex cells (sperm) are made, they need to have 23 chromosomes, one from each pair. The female gametes (eggs) also need 23 chromosomes. When they join at fertilisation, it will produce a cell called a **zygote** that has 46 chromosomes. This will grow into an embryo and then a baby.

This also means that the offspring that are produced from sexual reproduction are all different because they have different combinations of chromosomes from their mother and father.

Because the baby can receive any one of the 23 pairs from mum and any one from each of the 23 pairs from dad, the number of possible gene combinations is enormous.

This new mixture of genetic information produces a great deal of variation in the offspring.

This just mixes genes up in different combinations, but the only way that new genes can be made is by **mutation**. This is a random change in a gene.

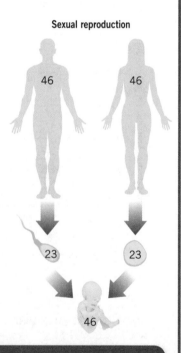

Sexual reproduction

46   46

23   23

46

✔ Maximise Your Marks

You only need to know the number of chromosomes in a human cell. Don't worry if a question asks about a different animal. Look for what information it supplies. For example, it might say that a sperm of a fruit fly has four chromosomes. You can then work out that a leg cell would have eight.

Sex Determination

In humans, the chromosomes of one of the 23 pairs are called the **sex chromosomes** because they carry the genes that determine the sex of the person.

There are two kinds of sex chromosome. One is called **X** and one is called **Y**:

- Females have two X chromosomes and are XX.
- Males have an X and a Y chromosome and are XY.

Females produce eggs that contain a single X chromosome. Males produce sperm, approximately half of which contain a Y chromosome and half of which contain an X chromosome.

This diagram shows the possible zygotes that can be produced by fertilisation.

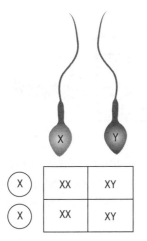

	X	Y
X	XX	XY
X	XX	XY

The reason why the sex chromosomes determine the sex of a person is due to a single gene on the Y chromosome. This gene causes the production of testes rather than ovaries and so the male sex hormone testosterone is made. This will cause the development of all the male characteristics.

Variation

Children born from the same parents all look slightly different. These differences are called **variation**. This can have different causes:

- **Inherited or genetic** – some variation is inherited from our parents in our genes.
- **Environmental** – some variation is a result of our environment.

Often our characteristics are a result of both our genes and our environment. The table shows examples of different kinds of variation.

Inherited	Environmental	Inherited and environmental
Earlobe shape	Scars	Intelligence
Eye colour	Spoken language	Body mass
Nose shape		Height
Dimples		

A good way to think of it is that our genes provide a height and weight range into which we will fit, and the amount we eat determines where in that range we will be.

Build Your Understanding

Scientists have argued for many years whether 'nature' or 'nurture' (inheritance or environment) is responsible for characteristics like intelligence, sporting ability and health.

Some of the most important work on this subject has been done by studying identical twins that have been separated at birth.

❓ Test Yourself

1. What does DNA stand for?
2. What does DNA code for?
3. Why is it important that a sex cell has only one chromosome from each pair?
4. What mechanism can produce new genes?

⭐ Stretch Yourself

1. Explain why approximately the same number of boys are born as girls.
2. Why are identical twins who have been separated at birth so useful when studying nature v nurture?

Passing on Genes

Different Copies of Genes

We have two copies of each chromosome in our cells (one from each parent). This therefore means that we have two copies of each gene.

Sometimes the two copies are the same, but sometimes they are different. A good example of this is tongue rolling. This is controlled by a single gene and there are two possible copies of the gene, one that says roll and the other that says do not roll. If a person has one copy of each then they can still roll their tongue.

This is because the copy for rolling is **dominant** and the non-rolling copy is **recessive**.

Each copy of a gene is called an **allele**. If both alleles for a gene are the same, this is called **homozygous**. **Heterozygous** means that the two alleles are different.

Genetic Screening

Genetic cross diagrams can only work out the probability of a child being affected.

It is now possible to test cells directly to see if they contain an allele for a particular genetic disorder. This is called **genetic screening**.

This can be done at different stages:
- In an **adult**. This can tell the person if they are a carrier for the disorder and therefore whether they can pass it on. It can also tell the person if he is going to develop a certain disorder later in life, for example Huntington's disease.
- In a **foetus**. Some cells can be taken from the foetus while the woman is pregnant. The parents can then find out if their baby will have the genetic disorder.
- In an **embryo** before it is implanted in the woman. If an embryo is produced by IVF outside the woman's body, then it can be tested before it is implanted in the woman. It is therefore possible to choose which embryos to put into the woman.

Genetic Crosses

We usually give the different copies (alleles) of a gene different letters, with the dominant copy a capital letter, for example T = tongue rolling and t = non-rolling.

Let us assume that mum cannot roll her tongue but dad can. Both of dad's alleles are T, so he is homozygous. This is called his **genotype** as it describes what alleles he has.

Rolling his tongue is called his **phenotype** as it describes the effect of the alleles.

The cross is usually drawn out like this:

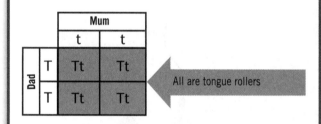

In this cross, all the children can roll their tongue. If both mum and dad are heterozygous, the children that they can produce may be different:

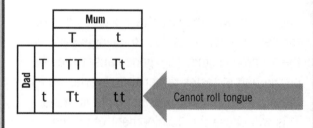

In this cross, 1 in 4 or 25% or a quarter of the children cannot roll their tongue.

✓ Maximise Your Marks

If you have to choose which letters to use in a cross, make sure that you use capital and small versions of the same letter and choose a letter that looks different in the two versions. Avoid *C* and *c* or *S* and *s*.

Genetic Disorders

Many genetic disorders are caused by certain copies of genes. These can be passed on from the mother or father to the baby and lead to the baby having the disorder. Examples of these disorders are cystic fibrosis, Huntington's disease and sickle-cell anaemia.

People with these disorders become ill, as shown in the table.

By looking at family trees of these genetic disorders and drawing genetic diagrams (such as the one for tongue rolling), it is possible for people to know the chance of their having a child with a genetic disorder. This may leave them with

a difficult decision to make as to whether to have children or not.

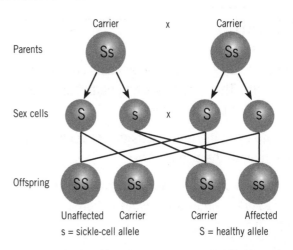

Cystic Fibrosis	Huntington's Disease	Sickle-Cell Anaemia
Caused by a recessive allele.	Caused by a dominant allele.	Caused by a recessive allele.
Symptoms include: • thick mucus collecting in the lung • difficulty in breathing • chest infections • food is not properly digested	Symptoms include: • muscle twitching (tremor) • loss of memory • difficulty in controlling movements • mood changes	Symptoms include: • feeling tired or weak • coldness in the hands and feet • pain in the bones, lungs and joints

Build Your Understanding

The process of genetic screening brings with it some difficult ethical decisions:
- As an adult, would you want to know if you were going to develop a disease for which there is no cure? Should your employer or your insurance company be told?
- In a foetus, the parents might have to decide whether to have a termination or not.
- In an embryo, the test is called preimplantation genetic diagnosis. Some people think that the destruction of early embryos is wrong. Others worry that the embryos may be tested and chosen for characteristics other than those involving disorders.

✓ Maximise Your Marks

To get an A* you must be able to describe arguments for and against genetic screening in a particular situation. Make sure you give both views.

? Test Yourself

1. Why can a person roll their tongue even if their cells have an allele for non-rolling?

2. Name a genetic disorder caused by a dominant allele.

3. What is the difference between genotype and phenotype?

4. What is genetic screening?

★ Stretch Yourself

1. What is the difference between a gene and an allele?

2. Suggest why people may not want their insurance company or employer to have the results of their genetic screening.

Introduction to Gene Technology

Cloning

Bacteria, plants and some animals can reproduce **asexually**. This only needs one parent and does not involve sex cells joining.

All the offspring that are made are genetically identical to the parent.

Gardeners often use **asexual reproduction** to copy plants – they know what the offspring will look like.

Different organisms have different ways of reproducing asexually:

- The spider plant grows new plantlets on the end of long shoots.
- Daffodil plants produce lots of smaller bulbs that can grow into new plants.
- Strawberry plants grow long runners that touch the ground and grow a new plant.

Asexual reproduction produces organisms that have the same genes as the parent.

Genetically identical individuals are called **clones**.

Many plants, such as the spider plant, clone themselves naturally and it is easy for a gardener to **take cuttings** to make identical plants.

Modern methods involve **tissue culture**, which uses small groups of cells taken from plants to grow new plants.

Cloning animals is much harder to do. Two main methods are used:

- **Cloning embryos** where embryos are split up at an early stage and the cells are put into host mothers to grow.
- **Cloning adult cells** – the first mammal to be cloned from adult cells was Dolly the sheep.

Since Dolly was born other animals have been cloned and there has been much interest about cloning humans.

There could be two possible reasons for cloning humans:

- **Reproductive cloning** to make embryos for infertile couples.
- **Therapeutic cloning** to produce embryos that can be used to treat diseases.

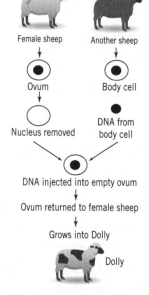

✓ Maximise Your Marks

Remember that clones have the same genes so any differences between them must be due to their environment.

Stem Cells

The use of embryos to treat disease is possible due to the discovery of **stem cells**.

Stem cells are cells that can divide to make all the different tissues in the body. They can be extracted from cloned embryos.

Scientists think that they could be used to repair damaged tissues such as injuries to the spinal cord.

There are therefore many different views about cloning.

Both infertility and genetic disease cause much pain and distress. I think that we should be able to use cloning to treat these problems.

It is not right to clone people because clones are not true individuals and it is not right to destroy embryos to supply stem cells.

Types of Stem Cell

There are two main types of stem cell:
- **Embryonic stem cells** can develop into any type of cell. It is easy to extract them from an embryo, but the embryo is destroyed as a result.

- **Adult stem cells** can develop into a limited range of cell types. It is not necessary to destroy an embryo to get them, but they are difficult to find.

Genetic Engineering

All living organisms use the same language of DNA. The four letters A, G, C and T are the same in all living things. Therefore a gene from one organism can be removed and placed in a totally

different organism where it will continue to carry out its function. This means, for example, that a cow can use a human gene to make the same protein as a human would make.

Moving a gene from one organism to another is called **genetic engineering**.

New **genetically modified (GM)** plants can be made in this way so that they:
- May be more resistant to insects eating them
- Can be resistant to herbicides (weedkillers).
- Can produce a higher yield.

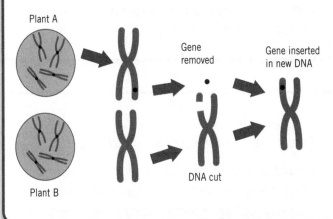

Plant A

Gene removed

Gene inserted in new DNA

DNA cut

Plant B

Build Your Understanding

People often have different views about GM crops.

Views against GM crops include:
- Genetic engineering is against 'God and Nature'.
- There may be long-term health problems with eating GM crops.
- Pollen from GM crops may spread to wild crops.

Views for GM crops include:
- There is more food for starving populations.
- They reduce need for harmful insecticides.

There is also the possibility that genetic engineering may be used to treat genetic disorders like cystic fibrosis. Scientists are trying to replace the genes in people that have the disorder with working genes.

Using genetic engineering to treat genetic disorders is called gene therapy.

❓ Test Yourself

1. Why does a gardener use cuttings rather than seeds to reproduce an attractive plant?

2. What was Dolly?

3. What can stem cells do that normal body cells cannot?

4. Write down one characteristic that is chosen for GM crops.

⭐ Stretch Yourself

1. Many people think that using adult stem cells to treat disease is acceptable, but are against using embryonic stem cells. Suggest why this is.

2. Suggest why farmers might want a crop that is resistant to herbicides.

Evolution and Natural Selection

Evolution

Most scientists now think that life on Earth started about 3500 million years ago.

How life started, and why there is such a great variety of organisms, are questions that people have argued over for a long time.

In the 1800s scientists started to find **fossils** of many different animals and plants. Many people at that time believed in creation. They said that God created organisms in their current state.

However, scientists found fossils of organisms such as dinosaurs that were no longer alive (extinct). Some people started to believe that species of organisms could gradually change.

Evolution is the gradual change in a species over a long period of time.

The problem for the believers in evolution was that at first they could not explain how the gradual changes happened.

Charles Darwin

Charles Darwin (1809–1882) was a naturalist on board a ship called the HMS Beagle. His job was to make a record of the wildlife seen at the places the ship visited.

On his travels, Darwin noticed four things:
- Organisms often produce **large numbers** of offspring.
- Population numbers usually remain constant over long time periods.
- Organisms are all slightly **different** – they show **variation**.
- This variation can be inherited from their parents.

Many of Darwin's observations were made on the Galapagos Islands off the coast of South America.

Darwin used these four simple observations to come up with a theory for how evolution could have happened.

Darwin said that:
- All organisms are slightly different.
- Some are better suited to the environment than others.
- These organisms are more likely to survive and reproduce.
- They will pass on these characteristics and over long periods of time the species will change.

Darwin called this theory **natural selection**.

Darwin was rather worried about publishing his ideas. When he finally published them they caused much controversy. Many people were very religious and believed in creation. It took many years before Darwin's theory was generally accepted.

Charles Darwin

✓ Maximise Your Marks

You need to be able to use Darwin's theory to explain how a group of organisms has evolved.

You may not have heard of the organisms before, but just use these main points:

variation ➡ **best adapted survive** ➡ **reproduce** ➡ **pass on genes**

Natural Selection in Action

Because natural selection takes a long time to produce changes, it is very difficult to see it happening. One of the first examples to be seen was the peppered moth.

This moth is usually light coloured, but after the Industrial Revolution a black type became common in polluted areas. This can be explained by natural selection.

Other examples that can be explained by natural selection include:
- Rats becoming resistant to the rat poison warfarin.
- Bacteria becoming resistant to antibiotics.

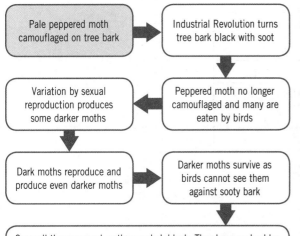

Pale peppered moth camouflaged on tree bark	Industrial Revolution turns tree bark black with soot
Variation by sexual reproduction produces some darker moths	Peppered moth no longer camouflaged and many are eaten by birds
Dark moths reproduce and produce even darker moths	Darker moths survive as birds cannot see them against sooty bark

Soon all the peppered moths are dark black. They have evolved by natural selection to suit their new environment

✓ Maximise Your Marks

When explaining how natural selection happens, remember to talk about groups or populations of organisms changing over time. One organism does not evolve; either it survives to reproduce or it dies.

Build Your Understanding

Darwin was not the first person to try to explain how evolution may have happened.

A French scientist called Lamarck (1744–1829) said that organisms were changed by their environment during their life. They then passed on the new characteristics and so the population changes.

Darwin's and Lamarck's ideas are theories that explain data known at the time. As different data becomes known then people often start to accept different theories.

Most scientists now think that Lamarck's theory is wrong because we now know that characteristics are passed on in our genes and genes are not usually altered by the environment.

Most people now accept Darwin's theory because it best explains all the data that has been discovered. But it is only a theory, it is not fact.

✓ Maximise Your Marks

To get an A* you must be able to spot an explanation of evolution made using Lamarck's ideas and explain why it is now thought to be wrong.

An example might be that giraffes have long necks because they stretched to reach higher leaves and in every generation they grew a little longer.

❓ Test Yourself

1. What does the word 'extinct' mean?
2. Write down one organism that has been investigated by looking at fossils.
3. What advantage did the dark moths have in industrial areas?
4. Why was Darwin worried about publishing his ideas?

⭐ Stretch Yourself

1. Use Darwin's theory of natural selection, rather than Lamarck's theory, to explain why giraffes have long necks.
2. The terms 'struggle for survival' and 'survival of the fittest' are often used when describing Darwin's ideas. Explain what they mean.

Practice Questions

 Complete these exam-style questions to test your understanding. Check your answers on page 122. You may wish to answer these questions on a separate piece of paper.

1 The theory put forward by Charles Darwin to explain how evolution could have occurred is called: (1)

   A   Selective Breeding

   B   Cloning

   C   Natural Selection

   D   Genetic Engineering

2 Which row **A**, **B**, **C** or **D** in the following table contains the correct characteristics? (2)

Characteristic is controlled by:			
	Genes Only	Environment Only	Both
A	Height	Eye colour	Body mass
B	Body mass	Scars	Earlobe shape
C	Scars	Body mass	Height
D	Blood group	Spoken language	Intelligence

3 A genetically identical copy of an individual is called a: (1)

   A   clone

   B   mutation

   C   gamete

   D   zygote

4 Finish the sentences by writing the correct word in the gap. (4)

   a)   The genes in cells are found in the part of the cell called the _____.

   b)   These genes are on long strands called _____.

   c)   They are made of a chemical called _____.

   d)   The genes control the cell by describing which _____ the cell should make.

5 The diagram shows the chromosomes present in a human skin cell.

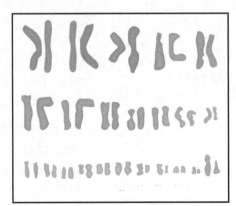

a) How would these chromosomes be different if they were from a sex cell and not a skin cell? (2)

....

....

b) The cell is from a male. How can you tell this? (2)

....

....

6 Jackie has just had a baby. Leroy is the father. The doctor has told them that their baby has cystic fibrosis even though neither Jackie nor Leroy have the disease.

a) Is the copy of the gene (allele) that causes cystic fibrosis dominant or recessive? How can you tell? (2)

b) Fill in the boxes below to show how Jackie and Leroy could have had a baby that has cystic fibrosis. Jackie's gametes have been done for you. (3)

	Jackie	
Gametes	F	f
Leroy		

c) If Jackie and Leroy had another baby, what would be the chance of it having cystic fibrosis? (1)

....

d) The doctor tells Jackie that if she gets pregnant again they could have the foetus tested to see if it has cystic fibrosis. Suggest why it may be a difficult decision for Jackie to decide whether to be tested or not. (2)

....

....

7 Rats can be killed using the poison warfarin. In the last 25 years many rat populations have become resistant to this poison. Use Darwin's theory of natural selection to explain how this has come about. (5)

....

....

....

....

....

How well did you do?

0–8 Try again   **9–14** Getting there   **15–20** Good work   **21–25** Excellent!

Classifying Organisms

Classifying Animals

Humans have been classifying organisms into groups ever since they started studying them.

- This makes it convenient when trying to identify an unknown organism.
- It also tells us something about how closely organisms are related and about their evolution.

The modern system that we use puts organisms into a system of smaller and smaller groups.

The groups are:

- kingdom
- phylum
- class
- order
- family
- genus
- species.

Kingdoms are the largest groups. The kingdoms are divided into smaller and smaller groups until the smallest group formed is called a **species**.

As you move down the groups there are fewer organisms in the group and they have more similarities.

💡 Boost Your Memory

There are lots of good ways of remembering the order of the groups from kingdom down to species. One example is **K**ing **P**hillip **c**ame **o**ver **f**or **g**reat **s**paghetti. You could always make up your own.

Difficulties with Species

Some organisms cause specific problems when trying to classify them as a species:

- Bacteria do not inter-breed, they reproduce asexually, so they cannot be classified into different species using the 'fertile offspring' idea.
- Hybrids are produced when members of two species inter-breed and so they are infertile. This occurs between many duck species.

Build Your Understanding

The characteristics that are used to classify organisms have changed over time.

The system used to be an artificial system based on one or two simple characteristics to make identification easier. One example was the presence of wings on an animal.

Now a natural system is used, which is based on evolutionary relationships. Animals that are more closely related are more likely to be in the same group.

To work out how closely related organisms are it is possible to study their DNA.

✓ Maximise Your Marks

To get an A* you must know that two organisms can have similar features for different reasons. They may be closely related, or they may be distantly related, but are both adapted for living in a similar environment.

Species

Members of a species are very similar, but how do we know if two similar animals are in the same species? Members of the same species can breed with each other to produce fertile offspring.

This means that horses and donkeys are different species because although they can mate and produce a mule, mules are infertile. The mule is an example of a **hybrid**.

Horse + donkey = mule!

Horse   Donkey   Mule

Different Groups

The first step in classifying an organism is to put it into a kingdom.

The five kingdoms are shown in the table:

Kingdom	Features
Prokaryotes (bacteria)	No nucleus
Animals	Multicellular, feed on other organisms
Plants	Cellulose cell wall, use light energy to produce food
Protoctista	Mostly single celled with some plant and some animal characteristics
Fungi	Cell wall of chitin, produce spores

Once an organism is put into the animal kingdom it can be put into the **vertebrate** phylum or one of several invertebrate phyla such as **arthropods**.

The vertebrates all have a backbone and the group is divided into five different classes:

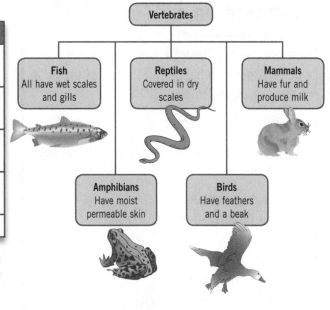

Vertebrates

Fish All have wet scales and gills

Reptiles Covered in dry scales

Mammals Have fur and produce milk

Amphibians Have moist permeable skin

Birds Have feathers and a beak

Naming Organisms

Organisms are often known by different names in different countries or even in different parts of the same country. All organisms are therefore given a scientific name by the international committees that are used by scientists in every country. This avoids confusion.

The scientific system of naming organisms is called the **binomial system**.

Each name has two parts. The first part is the name of the genus (the group above the species) The second part of the name is the species.

For example, the animals below are in the same genus but are different species. The genus starts with a capital letter but the species does not.

Lion is *Panthera leo*   Tiger is *Panthera tigris*

❓ Test Yourself

❶ An organism has cell walls made of chitin. What type of organism is it?

❷ What is a hybrid?

❸ Can tigers and lions mate to produce fertile offspring? Explain your answer.

❹ What are the characteristics of mammals?

⭐ Stretch Yourself

❶ Sharks are fish but dolphins are mammals. Why are they so similar in appearance?

❷ Archaeopteryx is an extinct animal. Fossils show that it had feathers and teeth. Why are there always going to be animals like archaeopteryx that are difficult to classify?

Competition and Adaptation

Competition

There are many different types of organism living together in a habitat and many of them are after the same things.

This struggle for resources is called **competition**.

The more similar the organisms, the greater the competition.

Plants usually compete for:
- light for photosynthesis
- water
- minerals.

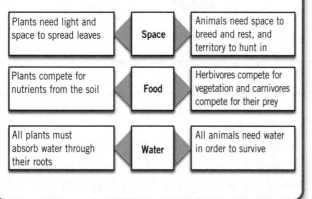

Plants need light and space to spread leaves	Space	Animals need space to breed and rest, and territory to hunt in
Plants compete for nutrients from the soil	Food	Herbivores compete for vegetation and carnivores compete for their prey
All plants must absorb water through their roots	Water	All animals need water in order to survive

Build Your Understanding

Organisms of the same species are more likely to compete with each other because they have similar needs.

A niche describes the habitat that an organism lives in and also its role in the habitat.

Organisms that share similar niches are more likely to compete with each other as they require similar resources.

There are different types of competition:
- Intraspecific is between organisms of the same species and is likely to be more significant as the organisms share more similarities and so need the same resources.
- Interspecific is between organisms of different species.

Adaptations to Extreme Conditions

Because there is constant competition between organisms, the best suited to living in the habitat survive.

Over many generations the organisms have became suited to their environment.

The features that make organisms well suited to their environment are called **adaptations**.

Habitats such as the Arctic and deserts are difficult places to live because of the extreme conditions found there. Animals and plants have to be well adapted to survive.

Polar bears have:	Cacti have:	Camels have:
A large surface area ratio to volume to minimise heat loss	Spines to reduce surface area to minimise water loss	A hump that stores food as fat
Thick insulating fur and a thick layer of fat under the skin	Water stored in the stem	Thick fur on top of the body for shade to protect the skin
White fur that is a poor radiator of heat and provides camouflage		Thin fur on the rest of the body to avoid overheating

✔ Maximise Your Marks

Be prepared to identify the adaptations on animals that you have not met before. Think about size, thickness of fur, and body fat.

Build Your Understanding

To prevent animals losing too much heat in cold climates they are usually quite large, like the polar bear, and have small ears. This helps to decrease the surface area to volume ratio.

All the members of a population may reproduce at the same time, so that predators would not be able to eat all the young. They may try to avoid the coldest temperatures by changing their behaviour.

Some animals will migrate long distances to warmer areas, while others may stay in the cold areas but slow down all their body processes and hibernate. When the sun is shining, animals like reptiles will lie in the sun or bask to try and increase their body temperature.

Some organisms, like polar bears, are very well adapted to living in specific habitats. These organisms are called specialists. They can survive in these areas when others cannot but would struggle to live elsewhere.

Other organisms, like rats, are not especially adapted to living in one habitat but can live in many areas. These organisms are called generalists. They will be outcompeted by specialists in certain habitats.

✓ Maximise Your Marks

To get an A* you must be able to explain why similar animals tend to be larger in Arctic regions and smaller in desert regions. Remember to talk about surface area to volume ratio.

Adaptations of Predators and Prey

Some animals called **predators** are adapted to hunt other animals for food.

The animals that are hunted are called **prey** and are adapted to help them escape.

Predators are adapted by having:
- Eyes at the front of their head which gives binocular vision to judge size and distance.
- Sharp teeth and claws to catch hold of prey.
- A body built for speed to chase prey.
- Stings or venoms (poison) to paralyse or poison prey.

Prey animals are adapted by having:
- A body that is camouflaged to avoid being seen by prey.
- Eyes on the side of head to give a view all around.
- A social organisation which involves living in groups which reduces the chance of being caught.
- A body built for speed to outrun predators.
- Defences such as stings or poison to deter predators eating them along with warning colouration.

? Test Yourself

1. Why do plants compete for light?
2. Why do camels have thick fur on the top of their body and thin fur underneath?
3. What is special about the leaves of cacti?
4. Why is it difficult to creep up behind a rabbit?

★ Stretch Yourself

1. Emperor penguins lay eggs and stay very close to the South Pole throughout the winter. Why does that make them unusual?
2. Elephants need to be large to digest vast quantities of poor quality food. What problem does this large size lead to in the desert and how does the elephant solve this?

Living Together

Predators and Prey

Organisms form different types of relationships with other organisms in their habitat. One of the most common is that of predator and prey.

The numbers of predators and prey in a habitat will vary and will affect each other.

The size of the two populations can be plotted on a graph that is usually called a predator–prey graph.

In the graph the peaks of the predator curve occur a little while after the peaks of the prey curve. This is because it takes a little while for the increase in food supply to allow more predators to survive and reproduce.

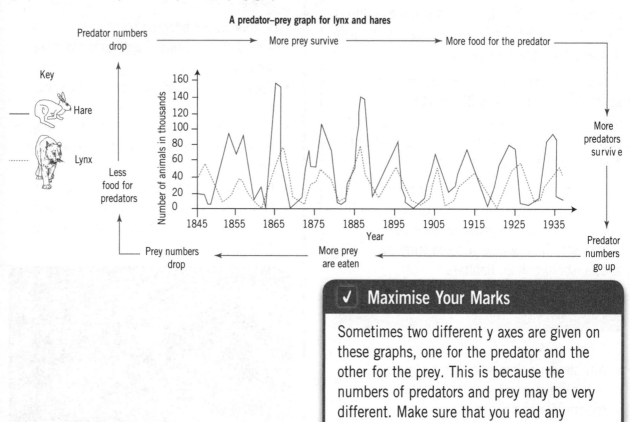

A predator–prey graph for lynx and hares

✓ Maximise Your Marks

Sometimes two different y axes are given on these graphs, one for the predator and the other for the prey. This is because the numbers of predators and prey may be very different. Make sure that you read any figures from the correct scale.

Mutualism

Instead of trying to eat each other, some different types of organisms work together.

When two organisms of different species work together so that both gain it is called **mutualism**.

Examples of this type of relationship are:
- Oxpeckers and buffaloes – the oxpecker birds eat the parasites on the backs of the buffalo, so the birds get food and the buffaloes get their parasites removed.
- Cleaner fish – these fish live in certain areas of the reef and are visited by larger fish. They do the same job as oxpeckers.
- Pollinating insects – they visit flowers and so transfer pollen allowing pollination to happen. They are 'rewarded' with sugary nectar from the flower.

Parasites and Hosts

Sometimes one organism may not kill another organism but it may take food from it while it is alive.

A **parasite** lives on, or in, another living organism called the **host**, causing it harm.

Many diseases, such as **malaria**, are caused by parasites feeding on a host.

The parasite in malaria is a single-celled species called Plasmodium that feeds on humans, who are the host. The organism is injected into the bloodstream by a mosquito. This is also acting as a parasite, but it is known as a vector for malaria because it spreads the disease-causing organism without being affected by it.

Mistletoe is a partial parasite. It grows on trees such as apple. It is green so it can photosynthesise and make its own food, but it also takes food from the apple tree.

Fleas are parasites that live in an animal's fur

Tapeworms are parasites that live in an animal's gut

Build Your Understanding

Pea plants and certain types of bacteria also benefit from mutualism.

Pea plants are legumes and have structures on their root called nodules. In these nodules nitrogen-fixing bacteria live.

The bacteria turn nitrogen gas into nitrogen-containing chemicals and give some to the pea plant.

The pea plant gives the bacteria some sugars that have been produced by photosynthesis.

Tube worms live deep in the ocean and cannot feed themselves. They have chemosynthetic bacteria living inside them. The bacteria can make their own food using the energy from chemical reactions. They give some of this to the worms in return for a safe place to live.

✔ Maximise Your Marks

To get an A* you must be able to link the presence of legumes with their nitrogen-fixing bacteria to the nitrogen cycle. This is discussed on page 53.

⚲ Boost Your Memory

Try making a list of all the organisms and relationships mentioned on pages 48–49, and then write down who gains and who suffers in each case.

❓ Test Yourself

1. When prey numbers are high then predator numbers start to increase. Why is this?
2. Why are fleas described as parasites?
3. Why is mistletoe called a partial parasite?
4. Why do flowers produce nectar?

★ Stretch Yourself

1. Lichens are mutualistic relationships. Algae grow inside the cells of fungi and get water and minerals from the fungi. Suggest what the fungi get in return.
2. Why is there so little food available deep in the oceans?

Energy Flow

Food Webs

A food web shows the feeding relationships between organisms in a habitat.

Each stage, or feeding level, in a food chain or food web is called a **trophic** level.

Producers are at the start of a food web because they can make their own food.

Most producers are green plants or algae that make food by photosynthesis.

Very few are bacteria, such as the ones that live in tube worms, who make food using energy from chemical reactions (chemosynthesis).

Very few organisms only eat one type of food. Most will eat several types and the food might be from different trophic levels.

For example, in this food web the birds eat both ladybirds and blackfly. When they eat blackfly they are secondary consumers and when they eat ladybirds they are tertiary consumers.

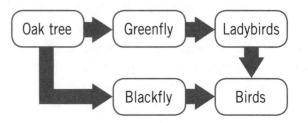

If one organism decreases or increases in numbers in a food web, it can alter the numbers of other organisms in the food web.

Pyramids of Biomass

The mass of all the organisms at each step of the food chain can be estimated.

This can be used to draw a diagram that is similar to a pyramid of numbers. The difference is that the area of each box represents the mass of all the organisms not the number.

This type of diagram is called a **pyramid of biomass**.

The reason that a pyramid of biomass is shaped like a pyramid is that energy is lost from the food chain in different ways as the food is passed along.

Often the waste from one food chain can be used by decomposers to start another chain.

The diagram shows that biomass and energy are lost from the food chain in a number of ways:

- In waste from the organisms by **excretion** and **egestion**.
- As heat when organisms **respire** – birds and mammals that keep a constant body temperature will often lose large amounts of energy in this way.

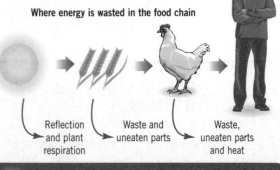

Where energy is wasted in the food chain

Reflection and plant respiration

Waste and uneaten parts

Waste, uneaten parts and heat

✓ Maximise Your Marks

Remember, excretion is the removal of waste products made by the body, e.g. urine, whereas egestion is food material that passes straight through. Candidates often get these two terms confused.

Build Your Understanding

Although pyramids of biomass are a better way of representing trophic levels, they are difficult to construct. This is because:

- Some organisms feed on organisms from different trophic levels.
- Measuring dry mass is difficult as it involves removing all the water from an organism, which will kill it.

The diagram below shows the flow of energy through a food chain.

Flow of energy through a food chain

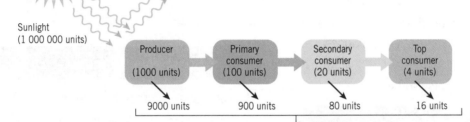

Of one million units of light that hit the surface of the plant, only 4 units are used for growth in the top consumer. This loss of energy also explains why food chains usually only have four or five steps. By then, there is so little energy left the animals would not be able to find enough food.

The percentage efficiency of transfer from producer to primary consumer is:

$$\frac{100}{1000} \times 100 = 10\%$$

This percentage is quite low because it is difficult to digest plant material.

✓ Maximise Your Marks

To get an A* you must be able to work out the percentage efficiency of energy transfer in different food chains and suggest reasons for differences. For example, farmers often keep their cattle indoors so that they lose less heat keeping warm.

? Test Yourself

1. Why are food webs more common than simple food chains in nature?

2. Look at the food web on page 50. What might happen to the numbers of ladybirds if a gardener killed all the blackfly in his garden with insecticide?

3. What does the size of a box represent in a pyramid of biomass?

4. Write down two ways that energy is lost from a food web.

★ Stretch Yourself

1. Calculate the percentage efficiency of the energy transfer from secondary to top consumer in the above food chain.

2. Compare your answer from question 1 with the percentage efficiency of transfer from producer to primary consumer. Suggest reasons for the difference.

Recycling

Decay

Some animals and plants die before they are eaten.

They also produce large amounts of waste products.

This waste material must be broken down or **decay** because it contains useful minerals. If this did not happen, organisms would run out of minerals.

Organisms that break down dead organic material are called **decomposers**. The main organisms that act as decomposers are bacteria and fungi.

They release enzymes on to the dead material that digest the large molecules. They then take up the soluble chemicals that are produced. The bacteria and fungi use the chemicals in respiration and for raw materials.

For decomposers to decay dead material they need certain conditions:

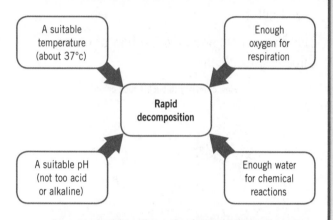

✓ Maximise Your Marks

Gardeners try to produce ideal conditions for decay in their compost heaps. Make sure you can explain how they do this.

⚲ Boost Your Memory

You need to remember the four conditions decomposers need to decay dead material – water, oxygen, temperature and the correct pH. Try to think of a way of remembering these four things. Perhaps make up a word that you can remember.

The Carbon Cycle

It is possible to follow the way in which each mineral element passes through living organisms and becomes available again for use. Scientists use nutrient cycles to show how these minerals are recycled in nature. One of these is the **carbon cycle**.

Carbon dioxide is returned to the air in a number of different ways:
- Plants and animals respiring.
- Soil bacteria and fungi acting as respiring decomposers.
- The burning of fossil fuels (combustion).

The main process that removes carbon dioxide from the air is photosynthesis.

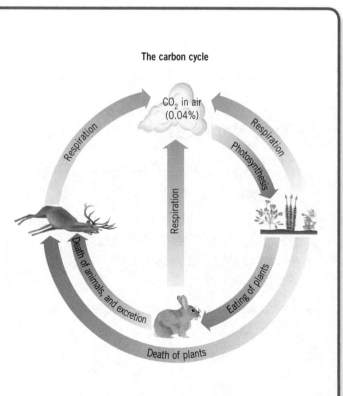

The carbon cycle

The Nitrogen Cycle

Plants take in nitrogen as **nitrates** from the soil to make protein for growth. Feeding passes nitrogen compounds along a food chain or web.

The nitrogen compounds in dead plants and animals are broken down by decomposers and returned to the soil.

The nitrogen cycle is more complicated than the carbon cycle because as well as the decomposers it involves three other types of bacteria:

- **Nitrifying bacteria** live in the soil and convert ammonium compounds to nitrates. They need oxygen to do this
- **Denitrifying bacteria** which live in the soil are the enemy of farmers. They turn nitrates into nitrogen gas. They need conditions without oxygen, rather than needing oxygen.
- **Nitrogen-fixing bacteria** they live in the soil or in special bumps called nodules on the roots of plants from the pea and bean family. They take nitrogen gas and convert it back to useful nitrogen compounds.

Although the air contains about 78% nitrogen, it is unreactive. This is why lightning and nitrogen-fixing bacteria are so important. They fix the nitrogen back into chemicals that can be used by plants.

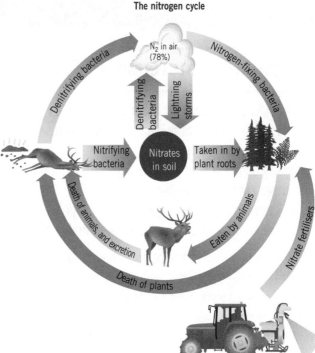

The nitrogen cycle

Denitrifying bacteria — N_2 in air (78%) — Nitrogen-fixing bacteria — Denitrifying bacteria — Lightning storms — Nitrifying bacteria — Nitrates in soil — Taken in by plant roots — Death of animals and excretion — Eaten by animals — Nitrate fertilisers — Death of plants

✓ Maximise Your Marks

To get an A* you must be able to recognise parts of the nitrogen cycle. You are unlikely to have to describe it all, but you might, for example, have to explain how the protein in fallen leaves gets converted to nitrates.

Build Your Understanding

Carbon dioxide is absorbed from the air by oceans. Much of this is by algae. Some marine organisms use the carbon dioxide and over millions of years make shells made of carbonate which become limestone rocks.

The carbon in limestone can return to the air as carbon dioxide during volcanic eruptions or weathering. The action of acid rain on buildings will speed up the weathering process due to the reaction with the limestone.

❓ Test Yourself

1. Gardeners water their compost heaps in dry weather. Why do they do this?

2. Write down two ways that carbon dioxide is returned to the atmosphere.

3. What is the main nitrogen-containing compound taken up by plants?

4. What do plants need nitrogen for?

⭐ Stretch Yourself

1. Explain why acid rain releases carbon dioxide when it falls on to certain buildings.

2. Farmers try to make sure that their soils are well drained so that they contain plenty of air. Explain why they do this.

Populations and Pollution

Population Increase

The human population on Earth has been increasing for a long time, but it is now going up more rapidly than ever:

- This is because of an increasing birth rate and decreasing death rate.
- The rate of increase of the population is increasing and this is called **exponential growth**.

This increase in the population is having a number of effects on the environment:

- More raw materials are being used up such as fossil fuels and minerals.
- More waste is being produced, which can lead to pollution.
- More land is being taken up to be used for activities such as building, quarrying, farming and dumping waste.

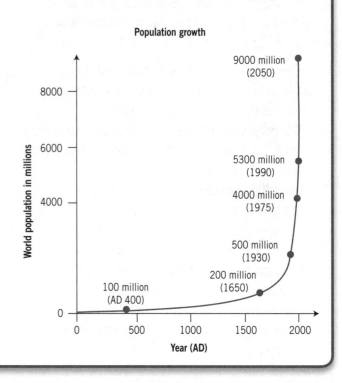

Population growth

Pollution

Modern methods of food production and the increasing demand for energy have caused many different types of **pollution**.

The table shows some of the main polluting substances that are being released into the environment.

Polluting Substance	Main source	Effects on the Environment
Carbon dioxide	Burning fossil fuels	Greenhouse effect
Sulfur dioxide	Burning fossil fuels	Acid rain
Chlorofluorocarbon (CFCs)	Fridges and aerosols	Destroys the ozone layer
Fertilisers including nitrates and phosphates	Intensive farming	Pollute rivers and lakes
Domestic waste	Households	Landfill sites release gases
Heavy metals	Industry	Accumulate in food chains
Sewage	Human and farm waste	Pollutes rivers and lakes

The **greenhouse effect** is caused by a build-up of gases, such as carbon dioxide and methane, in the atmosphere. These gases trap the heat rays as they are radiated from the Earth. This causes the Earth to warm up. This could lead to changes in the Earth's climate and a large rise in sea level.

Acid rain is caused by the burning of fossil fuels that contain sulfur impurities and nitrogen oxides. These give off sulfur dioxide, which dissolves in rainwater to form sulfuric acid. This falls as acid rain.

Ozone depletion is caused by the release of chemicals such as CFCs which come from the breakdown of refrigerators and aerosol sprays. Ozone helps protect us from harmful ultra-violet (UV) radiation, and so depletion may lead to more skin cancer.

Build Your Understanding

The world population figures show that the greatest rise in population is occurring in countries such as Africa and India.

However, the countries that use the most fossil fuels are developed countries, such as the USA and Europe.

A useful way of measuring how much pollution is caused per person is the carbon footprint. This measures the total greenhouse gas given off by a person or organisation over a certain period.

Rainwater containing fertilisers can run off from fields into rivers. Similarly, sewage also pollutes rivers. This can result in eutrophication of a river or lake:

- Nitrates and phosphates enter rivers and are absorbed by plants and algae. This makes them grow.

- Algae float near the water surface and their population increases dramatically. A 'blanket' of algae soon covers the surface.
- Other plants beneath the algae die as the surface algae block out the sunlight.
- Bacteria and other decomposers begin to break down the dead plants using more oxygen for respiration.

Fish die as the oxygen content of the water becomes too low.

✓ Maximise Your Marks

To get an A* you must be able to link the nitrogen cycle to eutrophication. Remember that nitrates in fertilisers are needed by plants to produce protein and are also produced when sewage is decomposed to ammonium compounds and then converted to nitrates by nitrifying bacteria.

Indicators of Pollution

Some organisms are more sensitive to pollution than others. If we look for these organisms it can tell us how polluted an area is.

On land, lichens grow on trees and stone. Some lichens are killed by lower levels of pollution than other types. Black spot fungus grows on roses in areas with less sulfur dioxide pollution.

In water, some animals, for example rat-tailed maggots, can live in polluted water, but other animals like mayfly nymphs can only live in clean water.

There are advantages to the different methods of measuring pollution:
- Using **indicator organisms** is cheaper and does not need equipment that can go wrong.
- Using **direct methods** can give more accurate results at any specific time.

Organisms and Environment

? Test Yourself

1 The world's population is showing an exponential increase. What does this mean?

2 Write down two gases that can cause the greenhouse effect.

3 What is the main gas that causes acid rain?

4 Write down one indicator organism that is found in polluted water.

★ Stretch Yourself

1 Explain why eating food that is grown in another country will result in a higher carbon footprint.

2 Nitrate fertilisers cause eutrophication. Explain how sewage can also cause eutrophication.

55

Conservation and Sustainability

Organisms at Risk

As well as causing pollution, the increasing demands for food, land and timber have caused people to cut down large areas of forests.

Deforestation has led to:
- Less carbon dioxide being removed from the air by the trees, and carbon dioxide being released when the wood is burnt.
- The destruction of habitats which contain rare species.

Some animals have been hunted until their numbers have been dramatically reduced, for example species of whales which have been hunted for food, oil and other substances. Their numbers are now very low and people are trying to protect them.

Other organisms have not been so lucky. Their numbers have decreased so much that they have completely died out. This is called **extinction**.

Organisms do become extinct naturally, but man has often increased the rate either directly or indirectly by:
- changing the climate
- over-hunting
- destroying habitats
- pollution
- competition.

Other organisms are at risk of becoming extinct and are **endangered**.

Biodiversity

Many people believe that it is wrong for humans to damage natural habitats and cause the death of animals and plants. They believe that it is important to keep a wide variety of different animals and plants alive.

The variety of different organisms that are living is called **biodiversity**.

There are many reasons given for trying to maintain biodiversity, such as:
- Losing organisms may have unexpected effects on the environment, such as the erosion caused by deforestation.
- Losing organisms may have effects on other organisms in their food web.
- Some organisms may prove to be useful in the future, such as for breeding, producing drugs or for their genes.
- Organisms may be needed for food.
- People enjoy looking at and studying different organisms.

Conservation Programmes

To help save and preserve habitats and organisms, people have set up many different **conservation** schemes.

There are a number of different ways that conservation programmes can work:

Make laws to make hunting illegal

Protect the habitats of organisms

Educate people about the importance of conservation

Breed endangered species in zoos and wildlife parks

Save the whale

Create artificial ecosystems

✓ Maximise Your Marks

Questions on this part of the course could use an animal or plant that you have never heard of. You may need to use the data that you are provided with to try to work out why it became endangered and what could be done to protect it.

Build Your Understanding

If populations become small then the variety of different alleles in the population might be quite low. The population has low genetic variation.

This makes it more likely to become extinct because:

- It will be harder for it to adapt to any changes in the environment because there is little variation.
- There is more chance of organisms being produced that have two identical harmful recessive alleles.

Reproducing with an organism that has similar alleles is called inbreeding. Zoos try to move animals to other zoos to breed with less related organisms to avoid this happening. It is also more likely to happen when populations become isolated.

✓ Maximise Your Marks

To get an A* you must be able to explain why it is so difficult to try to save endangered species once their numbers get below a certain level. Many studies indicate that about 500 individuals are needed to provide enough variation.

Sustainable Development

If the human population is going to continue to increase, it is important that we meet the demand for food and energy without causing pollution or over-exploitation.

Providing for the increasing population without using up resources or causing pollution is called **sustainable development**.

Fish stocks and woodland can be managed sustainably by:

- Educating people about the importance of controlling what they take from the environment.
- Putting quotas on fishing.
- Re-planting woodland when trees have been removed.

A decrease in the use of packaging materials and recycling would also help by:

- Cutting down on the energy needed to make them and transport them.
- Reducing the amounts of waste that has to be disposed of.

✓ Maximise Your Marks

Make sure that you can suggest why it is difficult to get people to agree to plans for sustainable development. Different people and different countries have different requirements. Some think that it is their right to hunt, fish or farm certain animals. Developing countries may not have enough money to provide alternatives.

? Test Yourself

1. Write down two ways that people have caused extinctions.
2. What is biodiversity?
3. What is conservation?
4. Why does recycling help to improve sustainability?

★ Stretch Yourself

1. Areas of tropical rainforest are being cleared, but small patches are being left. Explain why small isolated areas may not be very useful in conserving organisms.
2. About 10 000 years ago the cheetah nearly became extinct. Zoos find it very difficult to get cheetahs to produce healthy offspring. To make sure this happens zoos try to artificially inseminate cheetahs using sperm from around the world. Explain why they do this.

Practice Questions

Complete these exam-style questions to test your understanding. Check your answers on page 123. You may wish to answer these questions on a separate piece of paper.

1 Which of the following is **not** an effect of acid rain? (1)

A Lowers the pH of lakes. **C** Increases the temperature of the atmosphere.

B Poisons trees with aluminium salts. **D** Kills fish.

2 Certain fungi live on the roots of pine trees. The fungi take food from the tree roots. The fungi absorb minerals from the soil and some pass into the roots. The fungi is acting as a: (1)

A Parasite **C** Predator

B Mutualistic partner **D** Prey

3 The boxes below contain some types of animal and some descriptions. Draw straight lines to join each type of animal to the correct description. (3)

Type of Animal
Amphibian
Fish
Invertebrate
Mammal
Reptile

Description
Has wet scales.
Has moist, permeable skin.
Does not have a backbone.
Is covered in fur.
Has dry scales.

4 Organisms are adapted to the environment that they live in. Explain how the following characteristics help the organism survive.

a) Camels store large amounts of fat in their humps. (2)

b) Some cacti have deep roots that pass straight down, whereas other types of cacti have shallow roots that spread out over a long distance. (3)

c) Polar bears are large animals with very small ears for the size of their body. (2)

d) The larvae of many insects do not feed on the same type of food as the adult. (2)

5 A scientist estimated the numbers of greenfly and ladybirds in a large field. He estimated the numbers at different times of the year and plotted the results on a graph.

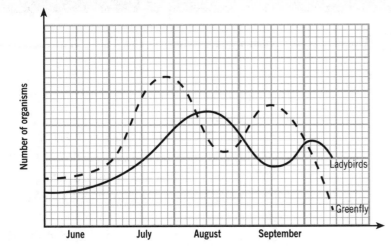

a) Explain the shape of the scientist's graph. (3)

b) What name is usually given to this type of graph? (1)

6 Read the following article carefully.

Protecting whales has often been one of the main aims of conservation groups. Records of whale hunting go back to 6000 BC. In the 1800s more modern methods of whaling reduced numbers dramatically and many whale species became endangered. Now an international group called the International Whaling Commission (IWC) tries to control whale hunting. Since 1985 commercial whaling has been banned, but some people disagree with the ban.

Japan still hunts whales for 'scientific research'. In the 2004–05 whaling season, 601 minke whales, three sperm whales and 51 Bryde's whales were taken. In 2005 the research programme increased the quota of minke whales to 900 and added fin whales to the programme. This move has sparked a great deal of controversy among anti-whaling nations because fin whales are listed as endangered.

a) Why is it difficult to be sure about the size of whale populations? (2)

b) Discuss arguments for and against the Japanese approach to whale hunting. (2)

How well did you do?

| 0–8 | Try again | 9–13 | Getting there | 14–17 | Good work | 18–22 | Excellent! |

Cells and Organisation

Plant and Animal Cells

Plants and animal cells have a number of structures in common.

They all have:
- A **nucleus** that carries genetic information and controls the cell.
- A **cell membrane** which controls the movement of substances in and out of the cell.
- **Cytoplasm** where most of the chemical reactions happen.

There are three main differences between plant and animal cells:
- Plant cells have a strong **cell wall** made of cellulose, whereas animal cells do not. The cell wall supports the cell and stops it bursting.
- Plant cells have a large permanent **vacuole** containing cell sap, but vacuoles in animal cells are small and temporary. The cell sap is under pressure and this supports the plant.
- Plant cells may contain **chloroplasts** containing chlorophyll for photosynthesis. Animal cells never contain chloroplasts.

Typical plant cell

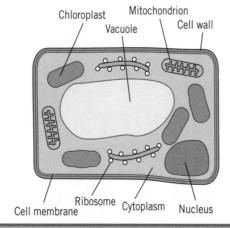

Typical animal cell

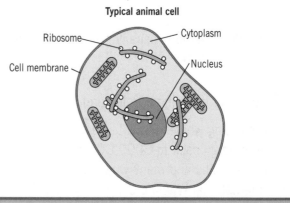

Build Your Understanding

The naked eye can see detail down to about 0.1 mm. Cells are smaller than this so a microscope is needed to see individual cells.

The best light microscopes can magnify cells so that objects as small as 0.002 mm can be seen clearly. At this magnification other structures in the cell become visible but cannot be seen clearly. Using an electron microscope allows objects as small as 0.000002 mm to be seen:
- Mitochondria are the site of respiration in the cell.
- Ribosomes are small structures in the cytoplasm where proteins are made.

Becoming larger and multicellular does have some advantages but also produces difficulties:
- It allows cells to specialise. This makes them more efficient at their job.

The difficulties that need to be solved are:
- It requires a communication system between cells to be developed.
- It is harder to supply all the cells with nutrients.
- The surface area to volume ratio is smaller so it is harder to exchange substances with the environment.

✓ Maximise Your Marks

To get an A* you must be able to measure cells and structures from a drawing and then use the magnification of the drawing to work out their size in real life. An example of this is question 1 in the Stretch Yourself section. Remember that magnification = image size ÷ size in real life.

Levels of Organisation

Some organisms are made of one cell. They are **unicellular**.

There seems to be a limit to the size of a single cell so larger organisms are made up of a number of cells. They are **multicellular**.

The cells are not all alike, but are specialised for particular jobs, for example guard cells in leaves and neurones in animals.

Similar cells, doing similar jobs, are gathered together into **tissues**, such as xylem and nerves. Different tissues are gathered together into **organs** to do a particular job, for example leaves and the brain.

Groups of organs often work together in **systems** to carry out related functions.

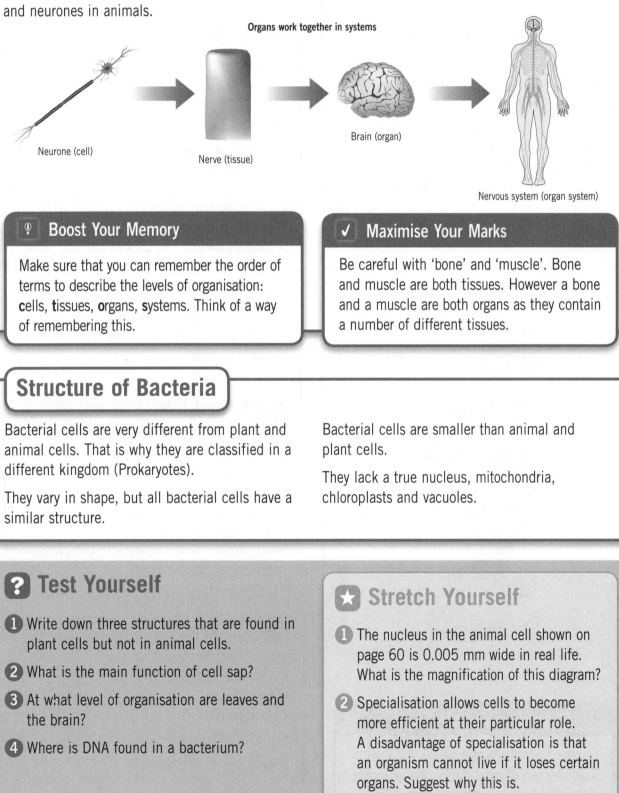

Organs work together in systems

Neurone (cell)

Nerve (tissue)

Brain (organ)

Nervous system (organ system)

💡 Boost Your Memory

Make sure that you can remember the order of terms to describe the levels of organisation: **c**ells, **t**issues, **o**rgans, **s**ystems. Think of a way of remembering this.

✔ Maximise Your Marks

Be careful with 'bone' and 'muscle'. Bone and muscle are both tissues. However a bone and a muscle are both organs as they contain a number of different tissues.

Structure of Bacteria

Bacterial cells are very different from plant and animal cells. That is why they are classified in a different kingdom (Prokaryotes).

They vary in shape, but all bacterial cells have a similar structure.

Bacterial cells are smaller than animal and plant cells.

They lack a true nucleus, mitochondria, chloroplasts and vacuoles.

❓ Test Yourself

1. Write down three structures that are found in plant cells but not in animal cells.
2. What is the main function of cell sap?
3. At what level of organisation are leaves and the brain?
4. Where is DNA found in a bacterium?

⭐ Stretch Yourself

1. The nucleus in the animal cell shown on page 60 is 0.005 mm wide in real life. What is the magnification of this diagram?
2. Specialisation allows cells to become more efficient at their particular role. A disadvantage of specialisation is that an organism cannot live if it loses certain organs. Suggest why this is.

DNA and Protein Synthesis

The Structure of DNA

The nucleus controls the chemical reactions occurring in the cell. This is because it contains the genetic material.

This is contained in structures called **chromosomes** which are made of **DNA**.

DNA is a large molecule with a very important structure:
- It has two stands.
- The stands are twisted to make a shape called a **double helix**.
- Each strand is made of a long chain of molecules including sugars, phosphates and bases.
- There are only four bases, called A, C, G and T.
- Links between the bases, hold the two chains together. C always links with G, and A with T.

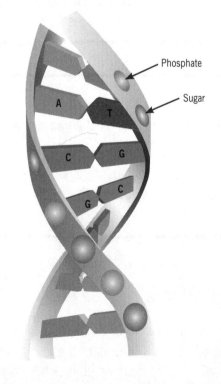

Phosphate

Sugar

Boost Your Memory

You need to remember **A** with **T** and **C** with **G**. Find a way of remembering it that means something to you. It could be **A**untie **T**ina and **C**ousin **G**eorge, for example.

Discovering the Structure of DNA

The two scientists that are famous for discovering the structure of DNA are Francis Crick and James Watson. They worked together in Cambridge in the early 1950s.

A molecule of DNA is about 0.00000034 mm wide, so they could not use a microscope to see it.

This is where they needed information obtained by other scientists:
- Maurice Wilkins and Rosalind Franklin fired X-rays at DNA crystals, and the images they obtained told Watson and Crick that DNA was shaped like a helix, with two chains.
- Erwin Chargaff had worked out that there was always the same percentage of the base C as G, and the same percentage of A as T.

These two pieces of information allowed Watson and Crick to build their famous model of the structure.

✓ Maximise Your Marks

How Science Works questions may ask for examples of how advances in science are made by cooperation between scientists. The discovery of the structure of DNA is a good example to use.

Build Your Understanding

DNA codes for the proteins it makes are called bases. There are four bases:
- adrenine (abbreviated A)
- cytosine (C)
- guanine (G)
- thymine (T)

They are given the letters A, C, G and T.

Coding for Proteins

DNA controls the cell by carrying the code for proteins:

- Each different protein is made of a particular order of amino acids, so DNA must code for this order.

- A gene is a length of DNA that codes for the order of amino acids in one protein.

Scientists now know that each amino acid in a protein is coded for by each set of three bases along the DNA molecule.

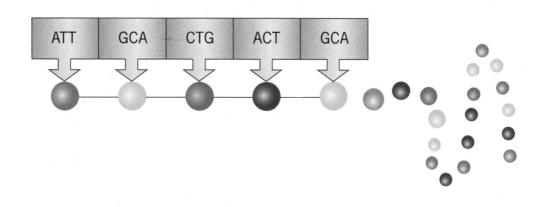

Build Your Understanding

Proteins are made on ribosomes in the cytoplasm and DNA is kept in the nucleus and cannot leave.

The cell has to use a messenger molecule to copy the message from DNA and carry the code to the ribosomes.

This molecule is called messenger RNA (mRNA).

✓ Maximise Your Marks

The word 'complementary' is a useful word to use when describing mRNA. It does not copy the code exactly but makes a version using the matching bases.

When a protein is to be made, these steps occur:
- The DNA containing that gene unwinds and 'unzips'.
- Complementary mRNA molecules pair up next to the DNA bases on one strand.
- The mRNA units join together and make a molecule with a complementary copy of the gene. This is called transcription.
- The mRNA molecule then leaves the nucleus and attaches to a ribosome.
- The base code on the mRNA is then used to link amino acids together in the correct order to produce the protein. Each three bases code for one amino acid. This is called translation.

❓ Test Yourself

1 Why is the shape of DNA described as a double helix?

2 How does the structure of one protein differ from the structure of another protein?

3 What holds the two strands together in a DNA molecule?

4 Why is the genetic code described as a triplet code?

⭐ Stretch Yourself

1 In a length of DNA, 34% of the bases are the base G. What percentage are base T?

2 When a protein is to be made, the length of DNA containing that gene 'unzips'. What does this mean and why is it necessary?

Proteins and Enzymes

The Functions of Proteins

The only way that the genetic material can control the cell is by coding for which proteins are made.

The proteins that are produced have a wide range of different functions:

- Structural proteins used to build cells, e.g. collagen.
- Hormones to carry messages, e.g. insulin.
- Carrier molecules, e.g. haemoglobin.
- Enzymes to speed up reactions, e.g. amylase.

Enzymes

Enzymes are biological catalysts.

Enzymes are produced in all living organisms and control all the chemical reactions that occur.

Most of the chemical reactions that occur in living organisms would occur too slowly without enzymes. Increased temperatures would speed up the reactions, but using enzymes means that the reactions are fast enough at 37°C.

These reactions include DNA replication, digestion, photosynthesis, respiration and protein synthesis.

✔ Maximise Your Marks

Many people think that all enzymes are released into the gut to digest food. Remember that most enzymes are found inside cells and are not released.

How do Enzymes Work?

As enzymes are protein molecules, they are made of a long chain of amino acids that is folded up to make a particular shape.

They have a slot or a groove, called the **active site**, into which the **substrate** fits. The substrate is the substance that is going to react.

The reaction then takes place and the **products** leave the enzyme. This explanation for how enzymes work is called the **Lock and Key theory**.

The substrate fits into the active site like a key fitting into a lock.

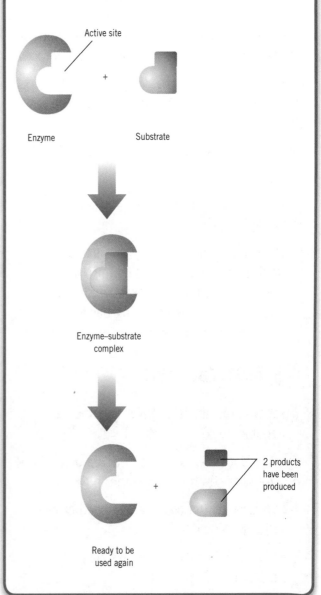

The lock and key theory of enzymes

Active site

Enzyme   +   Substrate

Enzyme–substrate complex

2 products have been produced

Ready to be used again

What Factors Affect Enzymes?

Enzymes work best at a particular temperature and pH. This is called the **optimum** temperature or pH.

Enzymes have different optimum values that depend on where they usually work.

If the concentration of the substrate is increased, then the reaction will be faster up to a certain concentration and then it will level off until all the enzymes are working at their maximum rate. At this point, increasing the substrate concentration does not increase the rate of reaction.

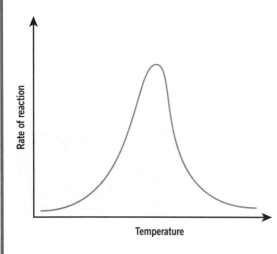

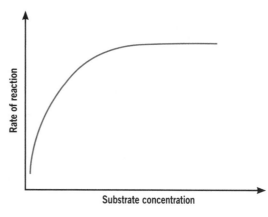

Build Your Understanding

The Lock and Key theory can be used to explain many of the properties of enzymes:

- It explains why an enzyme will only work on one type of substrate. They are described as specific. The substrate has to be the right shape to fit into the active site.
- If the temperature is too low, then the substrate and the enzyme molecules will not collide so often and the reaction will slow down. If the shape of the enzyme molecule changes, then the substrate will not easily fit into the active site. This means that the reaction will slow down.

High temperatures and extremes of pH may cause this to happen.

If the shape of the enzyme molecule is irreversibly changed, then it is described as being denatured.

✓ Maximise Your Marks

Many candidates lose marks by saying that heat kills enzymes. Remember that enzymes are protein molecules and not living organisms. Say that they are denatured or destroyed, but not killed.

❓ Test Yourself

1. Why does a lack of protein stunt growth?
2. Why are enzymes necessary in living organisms?
3. What is the lock and what is the key in the Lock and Key theory?
4. What does the phrase 'optimum temperature' mean?

⭐ Stretch Yourself

1. Lipase digests fats, but it will not digest proteins. Explain why this is.
2. Adding vinegar to food can stop the food being digested and spoilt by bacteria and fungi. Explain why this is.

Cell Division

Cells and Organisation

Copying the DNA

Before a cell divides, two things must happen. Firstly, new cell organelles such as mitochondria must be made. Secondly, the DNA must copy itself. Watson and Crick realised that the structure of DNA allows this to happen in a rather neat way:

- The double helix of DNA unwinds and the two strands come apart or 'unzip'.
- The bases on each strand attract their complementary bases and so two new molecules are made.

DNA replication

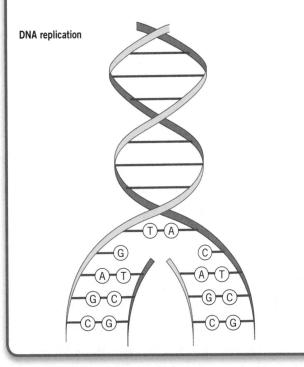

Types of Cell Division

Cells divide for a number of reasons. There are two types of cell division – **meiosis** and **mitosis** – and they are used for different reasons.

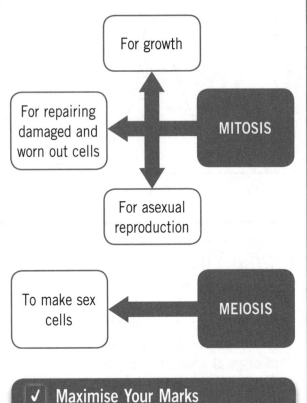

For growth

For repairing damaged and worn out cells

MITOSIS

For asexual reproduction

To make sex cells

MEIOSIS

✔ Maximise Your Marks

In most questions you will not lose marks if your spelling is a little inaccurate. However, make sure that you spell 'mitosis' and 'meiosis' correctly or the examiner may not know which one you mean.

Mitosis

In **mitosis**, two cells are produced from one. As long as the chromosomes have been copied exactly, then both new cells will have the same number of chromosomes and therefore the same genetic information as each other and the parent cell.

A cell has a pair of chromosomes that have divided

The two copies of each chromosome are separating

New nuclei are forming

The cytoplasm is dividing to make two cells each with two chromosomes

Meiosis (cont.)

In **meiosis**, the chromosomes are also copied once, but the cell divides twice. This makes four cells each with half the number of chromosomes, one from each pair.

Cells with one chromosome from each pair are called **haploid** and can be used as **gametes**. When two gametes join at fertilisation, the **diploid** or full number of chromosomes is produced.

| A cell has a pair of chromosomes each of which has divided | The two chromosomes are separating | Each double stranded chromosome is now split up | Four new cells are formed each with one chromosome |

Build Your Understanding

When DNA is copied, before mitosis and meiosis occur, mistakes are sometimes made.

A gene mutation occurs when one of the chemical bases in DNA is changed. This may mean that a different amino acid is coded for and this can change the protein that is made.

When this happens, it is most unlikely to benefit the organism. Either the protein will not be made at all or most likely it will not work properly.

Very occasionally, a mutation may be useful, and without mutations we would not be here.

Mutations occur randomly at a very low rate, but some factors can make them happen more often. These include:
- Ultra-violet light in sunlight.
- X-rays.
- Chemical mutagens as found in cigarettes.

Only mutations can produce new genes, but meiosis can recombine them in different orders.

Also, as a baby can receive any one of the chromosomes in each pair from the mother and any one from the father, the number of possible gene combinations is enormous.

This new mixture of genetic information produces a great deal of variation in the offspring.

This is why meiosis and sexual reproduction produces so much more variation than asexual reproduction.

✓ Maximise Your Marks

Remember that mitosis can produce cells that are genetically different, but this only happens if there is a mutation. Otherwise, they are genetically identical. Meiosis always produces genetic variation.

❓ Test Yourself

1. Does a new molecule of DNA have none, one or two original strands?
2. Where in the human body does meiosis occur?
3. The haploid number of chromosomes in humans is 23. What is the diploid number?
4. Write down two differences between mitosis and meiosis.

★ Stretch Yourself

1. Why is a gene mutation often harmful?
2. Why is it important to make sure that your sunglasses filter out UV light?

Growth and Development

Division and Differentiation

When gametes join at fertilisation, this produces a single cell called a **zygote**. The zygote soon starts to divide many times by mitosis to produce many identical cells.

These cells then start to become specialised for different jobs. The production of different types of cells for different jobs is called **differentiation**. These differentiated cells can then form tissues and organs.

Stem Cells

Some cells in the embryo and in the adult keep the ability to form other types of cells. They are called **stem cells**.

Scientists are now trying to use stem cells to replace cells that have stopped working or been damaged. This has the potential to cure a number of diseases.

Build Your Understanding

Once a cell has differentiated, it does not form other types of cell.

Although it has the same genes as all the other cells, many are turned off so it only makes the proteins it needs.

Scientists have found a way to switch genes back on and so have been able to clone animals from body cells. This is covered on page 38.

This means that it is now possible to produce embryos that are clones of an animal and to use them to supply embryonic stem cells.

There are many different views about the possibility of cloning humans to obtain stem cells.

A stem cell

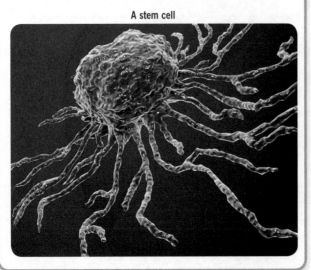

Human Growth Curves

Humans grow at different rates at different stages of their lives. This is shown in the graph.

The graph shows that there are two phases of rapid growth, one just after birth and the other in puberty.

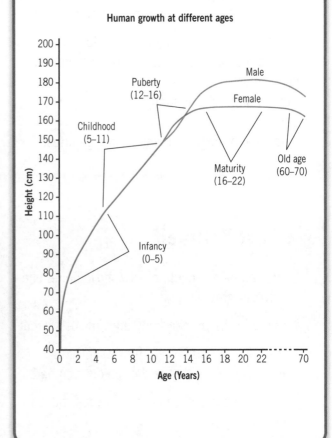

Human growth at different ages

Build Your Understanding

The various parts of the body also grow at different rates at different times.

faster while the brain and head grow more slowly into puberty and adulthood.

The diagram shows that the head and brain of an early foetus grow very quickly compared with the rest of the body.

Later, the body and legs start to grow

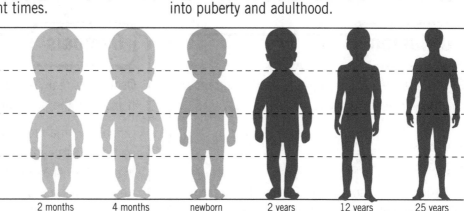

Plant Growth

Like animals, plants grow by making new cells by mitosis. The cells then differentiate into tissues like xylem and phloem. These tissues then form organs such as roots, leaves and flowers.

Growth in plants is different from animal growth in a number of ways:

- Plant cells enlarge much more than animal cells after they are produced. This increases the size of the plant.
- Cells tend to divide at the ends of roots and shoots. This means that plants grow from their tips.
- Animals usually stop growing when they reach a certain size, but plants carry on growing.
- Many plant cells keep the ability to produce new types of cells, but in animals only stem cells can do this. Plant cells that can produce new types of cells are called **meristematic**.

Measuring Growth

Growth can be measured as an increase in **height**, **wet mass** or **dry mass**. Dry mass is the best measure of growth. There are advantages and disadvantages of measuring growth by each method.

Measurement	Advantage	Disadvantage
Length or height	Easy to measure	Only measures growth in one direction
Wet mass	Fairly easy to measure	Water content can vary
Dry mass	Measures permanent growth over the whole body	Involves removing all the water from an organism

✓ Maximise Your Marks

It is easy to use the word 'weight' when talking about measuring growth, but you should really say 'wet mass' or 'dry mass'.

? Test Yourself

1. Write down one type of specialised cell.
2. What is a stem cell?
3. Look at the human growth curve. What does it show about growth in old age?
4. Which parts of a plant contain the main growth areas?

★ Stretch Yourself

1. Suggest why the head is much larger than the rest of the body when the baby is young?
2. Using dry mass to measure the growth of an organism presents a number of difficulties. Suggest what these difficulties are.

Transport in Cells

Cells and Organisation

Diffusion

Substances can pass across the cell membrane by three different processes:
- diffusion
- osmosis
- active transport.

Diffusion is the movement of a substance from an area of high concentration to an area of low concentration. Diffusion works because particles are always moving about in a random way. The rate of diffusion can be increased in a number of ways:

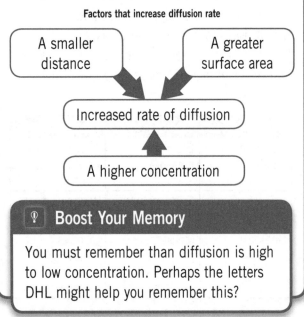

Factors that increase diffusion rate

A smaller distance → Increased rate of diffusion ← A greater surface area

A higher concentration → Increased rate of diffusion

💡 Boost Your Memory

You must remember than diffusion is high to low concentration. Perhaps the letters DHL might help you remember this?

Osmosis

Osmosis is really a special type of diffusion. It involves the diffusion of water.

Osmosis is the movement of water across a partially permeable membrane from an area of high water concentration to an area of low water concentration.

The cell membrane is an example of a partially permeable membrane. It lets small molecules through, such as water, but stops larger molecules, such as glucose.

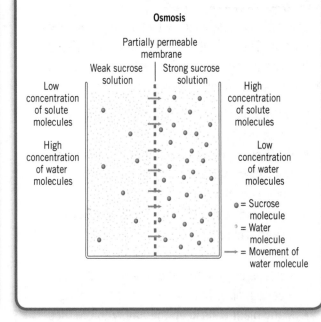

Osmosis

Partially permeable membrane

Weak sucrose solution | Strong sucrose solution

Low concentration of solute molecules

High concentration of water molecules

High concentration of solute molecules

Low concentration of water molecules

- = Sucrose molecule
- = Water molecule
→ = Movement of water molecule

Active Transport

Sometimes substances have to be moved from a place where they are in low concentration to one where they are in high concentration. This is in the opposite direction to diffusion and is called **active transport**.

Active transport is therefore the movement of a substance against the diffusion gradient with the use of energy from respiration.

Anything that stops respiration will therefore stop active transport. For example, plants take up minerals by active transport. Farmers try and make sure that their soil is not waterlogged because this reduces the oxygen content of the soil, so less oxygen is available to the root cells for respiration. This would therefore reduce the uptake of minerals.

✓ Maximise Your Marks

You can use the phrase 'up' or 'against the diffusion gradient' because this means in the opposite direction to diffusion. Don't say that active transport is 'along' or 'down the diffusion gradient' because this is the wrong way.

When plant cells gain water by osmosis, they swell. The cell wall stops them from bursting. This makes the cell stiff or turgid.

If a plant cell loses water it goes limp or flaccid.

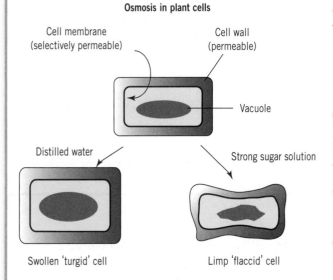

Osmosis in plant cells

Cell membrane (selectively permeable)

Cell wall (permeable)

Vacuole

Distilled water

Strong sugar solution

Swollen 'turgid' cell

Limp 'flaccid' cell

Turgid cells are very important in helping to support plants. Plants with flaccid cells often wilt.

Sometimes the cells lose so much water that the cell membrane may come away from the cell wall. This is called plasmolysis.

Animal cells do not behave in the same way because they do not have a cell wall.

They will either swell up and burst if they gain water, or shrink if they lose water.

It is possible to show how osmosis has occurred by cutting cylinders out of a potato and putting them into sugar solutions of different concentrations.

If the mass of the chips is measured before and after they are put in the solutions, a graph like this can be plotted:

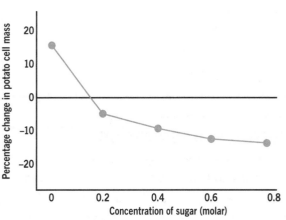

In less concentrated solutions, the potato gains water and increases in mass.

At high concentrations of sugar, the potato loses water and decreases in mass.

✓ Maximise Your Marks

Make sure you can explain the shape of this graph. It might be drawn as the length of the potato cylinder rather than the mass. It's just the same.

❓ Test Yourself

1. Why does the smell from a 'stink bomb' spread across a room?

2. Why do bad smells travel faster in warmer weather?

3. What is a partially permeable membrane?

4. What substance moves in osmosis?

★ Stretch Yourself

1. Look at the graph showing the percentage change in mass of the potato chips. What concentration of sugar solution is equal to the concentration inside a normal potato?

2. Why does a waterlogged soil have less oxygen in it?

Respiration

Cells and Organisation

What is energy needed for?

The energy that is released by respiration can be used for many processes:
- To make large molecules from smaller ones, for example proteins from amino acids.
- To contract the muscles.
- For mammals and birds to keep a constant temperature.
- For active transport.

Build Your Understanding

The energy released by respiration is needed for different processes in different parts of the cell. To make sure that the energy is not lost as heat, it is trapped in the bonds of a molecule called ATP.

ATP can then pass the energy on to wherever it is needed.

Aerobic Respiration

Aerobic respiration is when glucose reacts with oxygen to release energy. Carbon dioxide (CO_2) and water (H_2O) are released as waste products.

Glucose + Oxygen → Carbon + Water + **Energy**
                                dioxide

$$C_6H_{12}O_6 + 6O_2 \rightarrow 6CO_2 + 6H_2O + energy$$

The reactions of aerobic respiration take place in mitochondria.

All the reactions that occur in our body are called our **metabolism** and so anything that increases our **metabolic rate** will increase our need for respiration.

During exercise the body needs more energy and so the rate of respiration increases.

The breathing rate increases to obtain extra oxygen and remove carbon dioxide from the body. The heart beats faster so that the blood can transport the oxygen, glucose and carbon dioxide faster.

This is why our pulse rate increases.

Boost Your Memory

When you are learning the equation for respiration, look at the equation for photosynthesis in the next topic (page 78). Remember that one is just the reverse of the other. Don't try to learn them separately.

Anaerobic Respiration

When not enough oxygen is available, glucose can be broken down by **anaerobic respiration**.

This may happen in muscle cells during hard exercise.

In humans:
Glucose → Lactic acid + Energy

Being able to respire without oxygen sounds a great idea. However, there are two problems:

- Anaerobic respiration releases much less energy than is released by aerobic respiration.
- Anaerobic respiration produces lactic acid which causes muscle fatigue and pain.

In plants and fungi, such as yeast, anaerobic respiration is often called **fermentation**. It produces different products.

In plants and fungi:
Glucose → Ethanol + Carbon dioxide + Energy

Build Your Understanding

The extra oxygen required after exercise to deal with the build up lactic acid is called oxygen debt. Another name for this is excess post-exercise oxygen consumption (EPOC).

The lactic acid is transported to the liver and the heart continues to beat faster to supply the liver with the oxygen needed to break down the lactic acid.

It is possible to measure the respiration rate of organisms by measuring:
- The oxygen consumption.
- The carbon dioxide production.

This apparatus opposite can be used to investigate the rate of oxygen consumption by the maggots. If a liquid that absorbs carbon dioxide is placed in the bottom of the test tube, then the coloured liquid will move to the left.

It is then possible to investigate the effect of factors such as temperature or pH on the rate of respiration, for example by carrying out the experiment at different temperatures.

It is also possible to calculate the respiratory quotient (RQ) using this formula:

$$RQ = \frac{\text{carbon dioxide produced}}{\text{oxygen used}}$$

The RQ provides useful information about what type of substance is being respired.

Experiment to investigation the rate of oxygen consumption by maggots

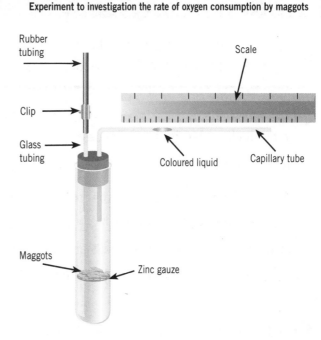

Cells and Organisation

✓ Maximise Your Marks

Remember that respiration is controlled by enzymes. This means that any factors that change the rate of enzyme reaction will also change the rate of respiration.

❓ Test Yourself

1. Why do we need to eat more in cold weather?
2. Why do we breathe faster when we exercise?
3. What are the bubbles of gas given off when yeast is fermenting glucose?
4. Why do our muscles hurt when we run a long race?

⭐ Stretch Yourself

1. Look at the equation for aerobic respiration using glucose. What would be the RQ of an organism that is respiring only glucose?
2. Wine makers need to carefully control the temperature inside the fermentation tanks when they make wine using fermentation. Explain why.

Practice Questions

 Complete these exam-style questions to test your understanding. Check your answers on page 123. You may wish to answer these questions on a separate piece of paper.

1 The following structures are found in plant and animal cells. Match words **A**, **B**, **C**, and **D**, with numbers **1–4** in the sentences. (4)

A mitochondria   **B** cell wall   **C** vacuole   **D** cell membrane

All organisms release energy from food. This largely happens in the _____**1**_____. Cells take up water by osmosis because the _____**2**_____ is partially permeable. The _____**3**_____ stores some sugars and salts. Plant cells are limited to how much water they can take up because the _____**4**_____ resists the uptake of too much water.

2 Complete the table putting a tick (✓) or a cross (✗) in the blank boxes. (4)

	Osmosis	Diffusion	Active Transport
Can cause a substance to enter a cell			
Needs energy from respiration			
Can move a substance against a concentration gradient			
Is responsible for oxygen moving into the red blood cells in the lungs			

3 The boxes contain some chemicals that are found in the cell and some functions. Draw straight lines to join each chemical to its correct function. (3)

Chemical
Amino acids
Bases
DNA
RNA

Function
Chromosomes are made from this chemical.
This is the chemical messenger that carries the genetic code out of the nucleus.
The order of these chemicals on the chromosomes codes for proteins.
These chemicals join together to make a protein molecule.

4 All the cells in the human body have about 20 000 genes. Scientists have studied some organs to see how many of these genes are used by cells in each organ. This number is shown below.

- Liver 2091
- Kidney 712
- Heart 1195
- Pancreas 1094
- Small intestine 297

a) Write down precisely where in a cell the genes are found. (2)

b) Genes are code for the production of proteins. Explain how each gene can code for a different protein. (2)

c) What percentage of its genes does each pancreas cell actually use? (1)

d) Which organ in the list would you expect to carry out the most chemical reactions? Explain your answer. (2)

5 Read the following article.

In 2006 scientists at Kyoto University performed an important experiment. They managed to turn body cells from rats into stem cells by inserting some genes. These genes seemed to reprogramme the cells so that they could use all their genes again. This experiment caused much excitement. This is because scientists hope that the technique can be used to produce stem cells in a way that fewer people object to.

Write about stem cells. In your answer include: (5)
- What they are.
- What they may be used for.
- Why this experiment might provide a solution to some people's objections.

How well did you do?

| 0–8 | Try again | 9–13 | Getting there | 14–18 | Good work | 19–23 | Excellent! |

Sampling Organisms

Where do Organisms Live?

Different organisms live in different environments.

- The place where an organism lives is called its **habitat**.
- All the organisms of one type living in a habitat are called a **population**.
- All the populations in a habitat are a **community**.
- An **ecosystem** is all the living and non-living things in a habitat.

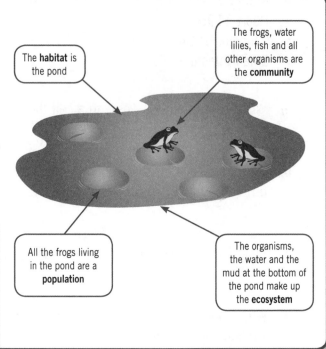

The **habitat** is the pond

The frogs, water lilies, fish and all other organisms are the **community**

All the frogs living in the pond are a **population**

The organisms, the water and the mud at the bottom of the pond make up the **ecosystem**

✔ Maximise Your Marks

Remember that for two organisms to be in the same population they must live in the same habitat and be in the same species. This means that they can successfully mate with each other.

Sampling an Area

It is possible to investigate where organisms live by using various devices.

A **quadrat** is usually a small square that is put on the ground within which all species of interest are noted or measurements taken. The number of organisms can be counted, or percentage cover estimated, in the quadrat and the size of the population in the whole area can then be estimated. Often several quadrats are required to determine the estimate.

It is easy to try to estimate how many of one type of plant live in a habitat:

- Work out the area of the whole habitat.
- Sample a small area using several quadrats and count out how many plants are present.
- Scale up this number to give an estimate of the whole habitat.

Quadrats are often used to study plants, but devices such as **pooters**, **nets** and **pit-fall traps** can be used to sample animal populations.

Working out the population of animals is harder because they do not keep still to be counted.

We can use a technique called **mark–recapture**:

- The organisms, such as snails, are captured, unharmed.
- They are counted and then marked in some way, for example the snail can be marked with a dot of non-toxic paint.
- They are released.
- Some time later the process of capturing is repeated and another count is made.
- This count includes the number of marked animals and the number of unmarked.

To work out the estimate of the population a formula is used. Population size is:

number in 1st sample × number in 2nd sample

number in 2nd sample previously marked

✔ Maximise Your Marks

Remember in all sampling questions, the more samples that you take in an area, the more accurate the estimate of the whole area will be.

Mapping an Area

To estimate the size of a population in an area we can use quadrats put down at random.

To see where the organisms live in a habitat we can use a **transect line**:

- A tape measure (or a piece of string) is put down in a line across the habitat.
- Quadrats are put down at set intervals along the tape.
- The organisms in the quadrats are then counted.

Artificial Ecosystems

Our planet has a range of different ecosystems. Some of these are **natural**, such as woodland and lakes. Others are **artificial** and have been created by man, such as fish farms, greenhouses and fields of crops.

Artificial ecosystems usually have less variety of organisms living there (less biodiversity). This may be caused by the use of chemicals such as weedkillers, pesticides and fertilisers.

Build Your Understanding

In some habitats a transect line can produce interesting results.

Different organisms live at different points along the line. This is because there is a change in the environmental conditions along the line. This is called zonation.

An example of zonation is found in a pond. Different plants can grow at different distances into the pond. This is due to the amount of water in the soil.

Trees

Bushes

Emergent vegetation

Floating vegetation

Submersed vegetation

❓ Test Yourself

1. What name is given to all the rabbits living in the same field?

2. What device would you use to sample:
   a) daisies in a field?
   b) butterflies?
   c) woodlice?

3. Five daisy plants are found in a 0.25 m² quadrat. How many would there be in a 100 m² field?

4. Why is it best to sample several areas in the field and take an average?

⭐ Stretch Yourself

1. A total of 30 snails are collected in an area, marked and released. When another sample is captured there are 35 snails and 5 are marked. What is an estimate of the snail population?

2. Different animals live on different parts of a rocky shore on the way down to the sea. Referring to tides, explain why the animals show zonation.

Photosynthesis

The Reactions of Photosynthesis

Plants make their own food by a process called **photosynthesis**. They take in carbon dioxide and water and turn them into sugars, releasing oxygen as a waste product. The process needs the energy from sunlight and this is trapped by the green pigment **chlorophyll**.

Carbon dioxide + Water → (Light / Chlorophyll) → Glucose + Oxygen

$$6CO_2 + 6H_2O \rightarrow C_6H_{12}O_6 + 6O_2$$

💡 Boost Your Memory

Try this for remembering the equation for photosynthesis: **C**ertain **w**orms **e**at **g**rass **o**utside (**c**arbon dioxide, **w**ater, **e**nergy, **g**lucose, **o**xygen).

Where does it happen?

Photosynthesis occurs mainly in the leaves.

The leaves are specially adapted for photosynthesis in a number of ways:
- A broad shape – provides a large surface area to absorb light and CO_2.
- A flat shape – the gases do not have too far to diffuse.
- Contain a network of veins – supply water from the roots and take away the products.
- Contain many chloroplasts in the palisade layer near the top – this traps the maximum amount of light.
- Pores called stomata (singular stoma) and air spaces – they allow gases to diffuse into the leaf and reach the cells.

Cross section of a leaf

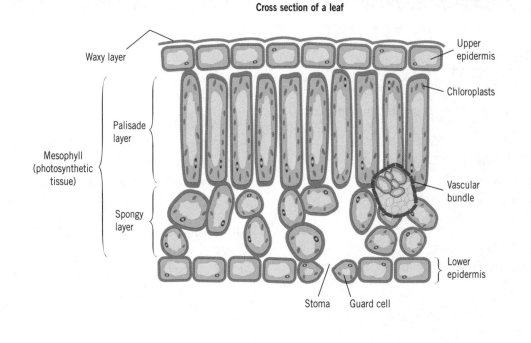

Waxy layer
Upper epidermis
Chloroplasts
Palisade layer
Mesophyll (photosynthetic tissue)
Spongy layer
Vascular bundle
Lower epidermis
Stoma
Guard cell

Photosynthesis Experiments

The understanding of the process of photosynthesis has changed considerably over time:

- Greek scientists thought that plants gained mass only by taking in minerals from the soil.
- Van Helmont in the 17th century worked out that plant growth could not be solely due to minerals from the soil. He found that the mass gained by a plant was more than the mass lost by the soil.
- In the 18th century Priestley showed that oxygen is produced by plants.

More modern experiments using isotopes have increased our understanding of photosynthesis. **Isotopes** of carbon can be used that behave in the same way chemically as carbon but which can be followed in the reactions because they are radioactive.

These experiments have shown that photosynthesis is a two-stage process:

- Light energy is used to split water, releasing oxygen gas and hydrogen atoms.
- Carbon dioxide gas combines with the hydrogen to make glucose and water.

Build Your Understanding

The rate of photosynthesis can be increased by providing:

- More light.
- More carbon dioxide.
- An optimum temperature.

Any of these factors can be limiting factors.

A limiting factor is something that controls how fast a reaction occurs. If more light is provided, it increases photosynthesis because more energy is available. After a certain point something else limits the rate.

Similarly more carbon dioxide increases the rate up to a point because more raw materials are present. Increasing the temperature makes enzymes work faster, but high temperatures denature enzymes.

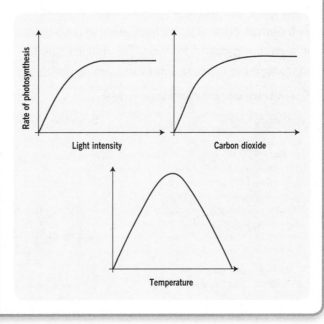

✓ Maximise Your Marks

Many candidates think that plants respire at night and photosynthesise during the day. To get an A* you must realise that plants carry out respiration all the time. Fortunately for us, during the day, they photosynthesise much faster than they respire, so overall release more oxygen than they take in.

? Test Yourself

1. What is the job of chlorophyll in photosynthesis?
2. Which cells in the leaf have most chloroplasts?
3. What are stomata?
4. Why do leaves have veins?

★ Stretch Yourself

1. In the reactions of photosynthesis, is oxygen released from water, carbon dioxide or both?
2. Why do high temperatures stop photosynthesis happening?

Food Production

Plants Need Minerals

Once plants have made sugars such as glucose by photosynthesis, they can convert it into many different things they need in order to grow:

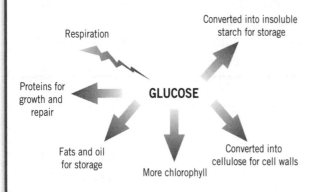

Respiration

Converted into insoluble starch for storage

Proteins for growth and repair

GLUCOSE

Fats and oil for storage

More chlorophyll

Converted into cellulose for cell walls

To produce these chemicals, plants need various minerals from the soil:

- **Nitrates** as a supply of nitrogen to make amino acids and proteins.
- **Phosphates** to supply phosphorus to make DNA and cell membranes.
- **Potassium** to help enzymes in respiration and photosynthesis.
- **Magnesium** to make chlorophyll.

Without these minerals plants do not grow properly. Farmers must therefore make sure that they are available in the soil.

Intensive Food Production

The human population is increasing and so there is a greater demand for food. This means that many farmers now use **intensive farming** methods.

Intensive farming means trying to obtain as much food as possible from the land. There are a number of different food production systems that use intensive methods:

Food production systems that use intensive methods

Fish farming

Fish are kept in enclosures away from predators. Their food supply and pests are controlled.

Glasshouses

Plants can be grown in areas where the climate would not be suitable. They can also produce crops at different times of the year.

Hydroponics

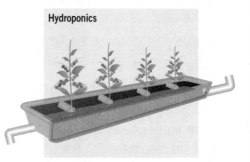

Plants are grown without soil. They need extra support but their mineral supply and pests are controlled.

Build Your Understanding

Farmers use a number of intensive farming techniques to help increase their yield but it is argued that the damage caused by some of these techniques does not justify the increase in food production:

- They use pesticides to kill pests that might eat the crop.

- They use herbicides to kill weeds that would compete with the crop.
- They keep animals indoors so that they do not waste energy keeping warm or moving about.
- They provide the plants with chemical fertilisers for growth.

Organic Food Production

Many people think that intensive farming is harmful to the environment and cruel to animals. Farming that does not use the intensive methods is called **organic farming**. Organic farming uses a number of different techniques:

Technique	Details
Use of manure and compost	These are used instead of chemical fertilisers and provide minerals for the plant
Crop rotation	Farmers do not grow the same crop in the same field year after year; this stops the build-up of pests and can reduce nutrient depletion of the soil
Use of nitrogen-fixing crops	These crops contain bacteria that add minerals to the soil
Weeding	This means that chemical herbicides are not needed
Varying planting times	This can help to avoid times that pests are active
Using biological control	Farmers can use living organisms to help to control pests; the organisms may eat the pests or cause disease

Preserving Food

Preservation method	How it is done	How it works
Canning	Food is heated in a can and the can is sealed	The high temperature kills the microorganisms, and oxygen cannot get into the can after it is sealed
Cooling	Food is kept in a refrigerator at about 5°C	The growth and respiration of the decomposers slow down at low temperature
Freezing	Food is kept in a freezer at about −18°C	The decomposers cannot respire or reproduce
Drying	Dry air is passed over the food	Microorganisms cannot respire or reproduce without water
Adding salt or sugar	Food is soaked in a sugar solution or packed in salt	The sugar or salt draws water out of the decomposers
Adding vinegar	The food is soaked in vinegar	The vinegar is too acidic for the decomposers

Although gardeners want decay to happen in their compost heaps, people do not want their food to decay before they can eat it. **Food preservation** methods reduce the rate of decay of foods. There are many ways to preserve food. Most stop decay by taking away one of the factors that decomposers need.

✓ **Maximise Your Marks**

Look back at page 52 which covers decay. Make sure that you can explain how a food preservation method stops decay.

❓ Test Yourself

1 Why do plants need nitrates?

2 Why does a plant look yellow if grown with a lack of magnesium?

3 What is hydroponics?

4 Why does food still go bad in a refrigerator?

⭐ Stretch Yourself

1 Suggest one problem with using large quantities of chemical pesticides to kill insect pests.

2 In intensive farming, why is the food brought to pigs rather than letting them find food?

Ecology

Practice Questions

Complete these exam-style questions to test your understanding. Check your answers on page 124. You may wish to answer these questions on a separate piece of paper.

1 Put a cross (**✗**) next to any farming method that would **not** be used by organic farmers.　　(2)

Spreading manure on the fields. ☐　　　　Killing weeds using weedkillers. ☐

Spraying chemical pesticides. ☐　　　　Rotating their crops. ☐

2 Arthur wants to measure how fast a plant photosynthesises at different light intensities. The diagram shows the apparatus he uses.

Arthur includes the following steps in his method:

- He uses the same piece of pondweed for the complete investigation.
- He adds sodium hydrogen carbonate to the water to provide the plant with carbon dioxide.
- He times five minutes using the minute hand of his watch and counts the number of bubbles given off.
- He counts the bubbles three different times at each different light intensity.
- He repeats this with the light at different distances from the pondweed.

a) Why does Arthur choose pondweed for his experiment?　　(2)

b) What is the main gas found in the bubbles?　　(1)

c) Write down the step that helps to make Arthur's experiment valid.　　(1)

d) Suggest one way that Arthur could make his experiment more accurate.　　(1)

3 Two pupils were studying a rocky shore next to the sea. They drew the shape of the shore.

The pupils then measured the distribution of one type of seaweed down the shore.

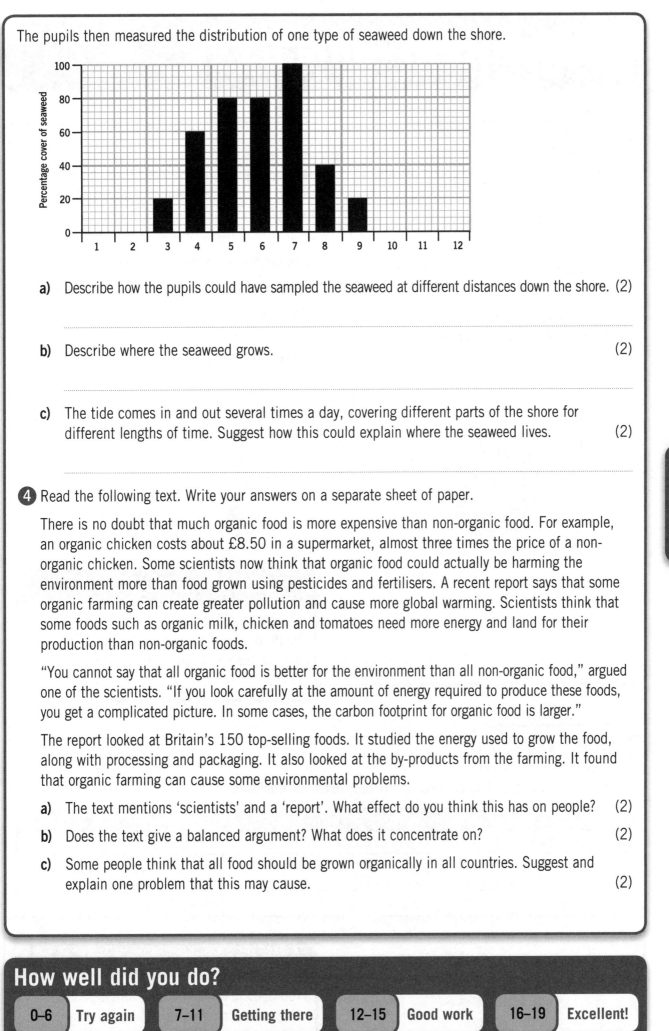

a) Describe how the pupils could have sampled the seaweed at different distances down the shore. (2)

....

b) Describe where the seaweed grows. (2)

....

c) The tide comes in and out several times a day, covering different parts of the shore for different lengths of time. Suggest how this could explain where the seaweed lives. (2)

....

4 Read the following text. Write your answers on a separate sheet of paper.

There is no doubt that much organic food is more expensive than non-organic food. For example, an organic chicken costs about £8.50 in a supermarket, almost three times the price of a non-organic chicken. Some scientists now think that organic food could actually be harming the environment more than food grown using pesticides and fertilisers. A recent report says that some organic farming can create greater pollution and cause more global warming. Scientists think that some foods such as organic milk, chicken and tomatoes need more energy and land for their production than non-organic foods.

"You cannot say that all organic food is better for the environment than all non-organic food," argued one of the scientists. "If you look carefully at the amount of energy required to produce these foods, you get a complicated picture. In some cases, the carbon footprint for organic food is larger."

The report looked at Britain's 150 top-selling foods. It studied the energy used to grow the food, along with processing and packaging. It also looked at the by-products from the farming. It found that organic farming can cause some environmental problems.

a) The text mentions 'scientists' and a 'report'. What effect do you think this has on people? (2)

b) Does the text give a balanced argument? What does it concentrate on? (2)

c) Some people think that all food should be grown organically in all countries. Suggest and explain one problem that this may cause. (2)

How well did you do?

0–6 Try again   **7–11** Getting there   **12–15** Good work   **16–19** Excellent!

Transport in Animals

Physiology

Blood

Blood is made up of a liquid called **plasma**.

Plasma carries chemicals such as dissolved food, hormones, antibodies and waste products around the body.

Cells are also carried in the plasma. They are adapted for different jobs.

Red blood cells are shaped like a biconcave disc. They contain haemoglobin which can carry oxygen around the body.

The haemoglobin in red blood cells reacts with oxygen in the lungs, forming oxyhaemoglobin. In the tissues, the reverse of this reaction happens and oxygen is released:

haemoglobin + oxygen $\rightleftharpoons$ oxyhaemoglobin

White blood cells can change shape to engulf and destroy disease causing organisms. They can also produce antibodies.

Platelets are responsible for clotting the blood.

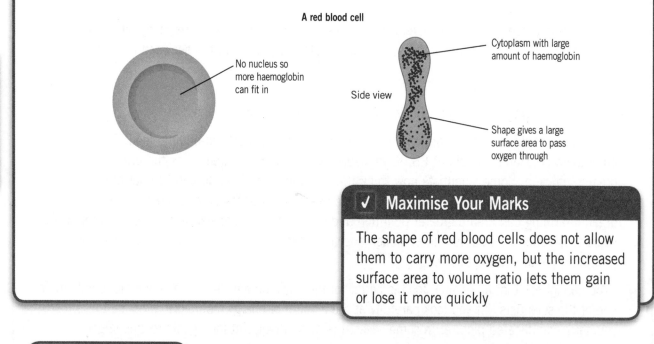

A red blood cell

No nucleus so more haemoglobin can fit in

Side view

Cytoplasm with large amount of haemoglobin

Shape gives a large surface area to pass oxygen through

✓ Maximise Your Marks

The shape of red blood cells does not allow them to carry more oxygen, but the increased surface area to volume ratio lets them gain or lose it more quickly

Blood Vessels

The blood is carried around the body in **arteries**, **veins** and **capillaries**.

Arteries	Veins	Capillaries
Carry blood away from the heart	Carry blood back to the heart	Join arteries to veins
Have thick, muscular walls because the blood is under high pressure	Have valves and a wide lumen because the blood is under low pressure	Have permeable walls so that substances can pass in to and out of the tissues

💡 Boost Your Memory

You need to remember that **a**rteries carry blood **a**way from the heart and ve**in**s carry it back **in**to the heart.

The Heart

The heart is made up of four chambers.

The top two chambers are called **atria** and they receive blood from veins.

The bottom two chambers are **ventricles**. They pump the blood out into arteries.

The top two chambers, the atria, fill up with blood returning in the **vena cavae** and **pulmonary veins**. The two atria then contract together and pump the blood down into the ventricles. The two ventricles then contract, pumping blood out into the **aorta** and **pulmonary arteries** at high pressure.

In the heart are two sets of valves, whose function is to prevent blood flowing backwards.

In between the atria and the ventricles are the **bicuspid** and **tricuspid valves**.

These valves stop blood flowing back into the atria when the ventricles contract. The pressure of blood closes the flaps of the valves and the tendons stop the flaps turning inside out.

There are also **semi-lunar** valves between the ventricles and the arteries.

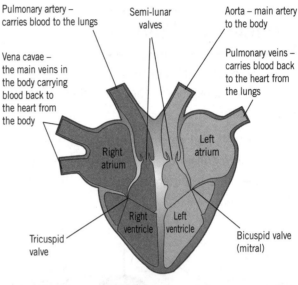

Cross section of a heart

Pulmonary artery – carries blood to the lungs

Semi-lunar valves

Aorta – main artery to the body

Vena cavae – the main veins in the body carrying blood back to the heart from the body

Pulmonary veins – carries blood back to the heart from the lungs

Right atrium

Left atrium

Right ventricle

Left ventricle

Tricuspid valve

Bicuspid valve (mitral)

Physiology

✓ Maximise Your Marks

Make sure that you can spot that the muscle wall of the left ventricle is always thicker than that of the right ventricle. This is because it has to pump blood all round the body compared with the short distance to the lungs.

Build Your Understanding

Mammals have a double circulation.

This means that the blood has to pass through the heart twice on each circuit of the body.

Deoxygenated blood is pumped to the lungs and the oxygenated blood returns to the heart to be pumped to the body.

The advantage of this system is that the pressure of the blood stays quite high and so it can flow faster around the body.

Because of the double circulation the heart is really two pumps in one:
* The right side pumps the blood to the lungs.
* The left side pumps it to the rest of the body.

❓ Test Yourself

1. What is the job of platelets?
2. Why do red blood cells lack a nucleus?
3. Why do veins have valves?
4. What blood vessel carries blood from the heart to the lungs?

⭐ Stretch Yourself

1. Why is the right side of the heart coloured blue in the diagram?
2. Some people have a defect in the bicuspid valve. Explain why this can lead to a build up of blood in the blood vessels of the lungs.

Transport in Plants

Xylem and Phloem

Plants have two different tissues that are used to transport substances. They are called **xylem** and **phloem**.

Xylem vessels and phloem tubes are gathered together into collections called **vascular bundles**. They are found in different regions of the leaf, stem and root.

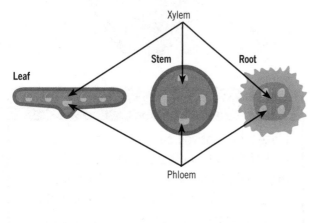

Xylem vessels and phloem tubes are different in structure and do different jobs:

Xylem	Phloem
Carries water and minerals from roots to the leaves	Carries dissolved food substances both up and down the plant
The movement of water up the plant and out of the leaves is called transpiration	The movement of the dissolved food is called translocation
Made of vessels which are hollow tubes made of thickened dead cells	Made of columns of living cells

💡 Boost Your Memory

Remember: phloem for food and xylem for water.

The Movement of Water

Water enters the plant through the **root hairs** by **osmosis**.

The root hair cells increase the surface area for the absorption of water.

Water then passes from cell to cell by osmosis until it reaches the centre of the root.

The water enters xylem vessels in the root and then travels up the stem.

Water enters the leaves and evaporates.

It then passes through the **stomata** by **diffusion**.

This loss of water is called **transpiration** and it helps to pull water up the xylem vessels.

Various environmental conditions can affect the transpiration rate.

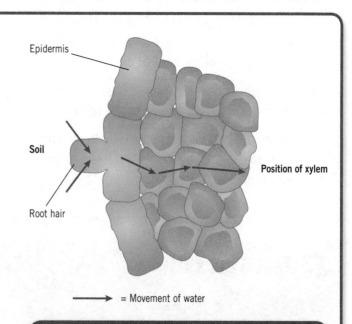

✓ Maximise Your Marks

Remember that it is osmosis that brings the water into the leaf and into the xylem, but water does not move up the xylem by osmosis. It is 'sucked up' by evaporation from the leaves.

Transpiration Rate

The rate of transpiration depends on a number of factors:

- **Temperature** – warm weather increases the kinetic energy of the water molecules so they move out of the leaf faster.
- **Humidity** – damp air reduces the concentration gradient so the water molecules leave the leaf more slowly.
- **Wind** – the wind blows away the water molecules so that a large diffusion gradient is maintained.
- **Light** – light causes the stomata to open and so more water is lost.

The factors that speed up transpiration will also increase the rate of water uptake from the soil. If water is scarce, or the plant roots are damaged, the plants chances of survival is increased if transpiration can be slowed down. Plants do this by wilting, or can be cut so that they can grow new roots.

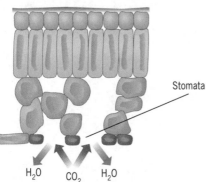

Stomata

H_2O CO_2 H_2O

✔ Maximise Your Marks

If a question asks you to 'give one factor that increases transpiration rate', make sure that you write 'an increase in temperature', not just 'temperature'. Many candidates lose marks in this way.

Build Your Understanding

When plants are short of water, they do not want to waste it through transpiration. The trouble is they need to let carbon dioxide in, so water will always be able to get out. Water loss is kept as low as possible in several ways:

- Photosynthesis only occurs during the day, so the stomata close at night to reduce water loss. The guard cells lose water by osmosis and become flaccid. This closes the pores.
- The stomata are on the underside of the leaf. This reduces water loss because they are away from direct sunlight and protected from the wind.
- The top surface of the leaf, facing the Sun, is often covered with a protective waxy layer.

Although transpiration is kept as low as possible, it does help plants by cooling them down and supplying leaves with minerals. It also provides water for support and photosynthesis.

Physiology

? Test Yourself

1. In which direction in a plant stem do water and minerals move?
2. What is translocation?
3. Where in a plant root is xylem found?
4. What is the function of root hair cells?

★ Stretch Yourself

1. Why is it impossible for plants to prevent all water loss from the leaves?
2. What causes stomata to close when a plant wilts?

Digestion and Absorption

Digestion

The job of the digestive system is to break down large food molecules into small soluble molecules. This is called **digestion**.

Digestion happens in two main ways – **physical** and **chemical** digestion.

Physical digestion occurs in the mouth where the teeth break up the food into smaller pieces.

Chemical digestion is caused by digestive enzymes that are released at various points along the digestive system. Most enzymes work inside cells, controlling reactions. Some enzymes pass out of cells and work in the digestive system. These enzymes digest our food, making the molecules small enough to be absorbed.

The food is moved along the gut by contractions of the muscle in the lining of the intestine. This process is called **peristalsis**.

The digestive system

Saliva is released into the mouth from the salivary glands. It contains amylase to break down starch to maltose.

The liver makes bile that contains bile salts. They break the large fat droplets down into smaller droplets. Bile is stored in the gall bladder.

The small intestine makes enzymes such as maltase. This breaks down maltose to glucose.

The stomach makes gastric juice, containing protease and hydrochloric acid. The acid kills microbes and creates the best pH for the protease to digest proteins.

The pancreas makes more protease and amylase. It also makes lipase to break down the fats to fatty acids and glycerol.

Build Your Understanding

To make the digestive enzymes work at an optimum rate, the digestive system provides the best conditions:

* Each enzyme has a different optimum pH. Protease in the stomach works best at about pH 2, but a different protease made by the pancreas works best at about pH 9.

* Physical digestion helps to break the food into smaller particles, thereby increasing the surface area of the food particles. Bile salts emulsify fat droplets, breaking them into smaller droplets so lipase can work faster.

✓ Maximise Your Marks

Be careful not to say that bile salts break down fats. Make sure that you say 'into fat droplets', otherwise it sound like bile salts are doing the same job as lipase.

Absorption

In the small intestine, small digested food molecules are absorbed into the bloodstream by diffusion. The inside of the small intestine is permeable and has a large surface area over which absorption can take place.

The lining of the small intestine contains two types of vessel that absorb the products of digestion:
- **Capillaries** absorb food and take it to the liver via the **hepatic portal vein**.
- **Lacteals** absorb mainly the products of fat digestion and empty them into the bloodstream.

Build Your Understanding

A number of factors increase the surface area of the small intestine and so speed up the rate of absorption:
- The human small intestine is over 5 metres long.
- The inner lining is folded.
- The folds are covered with finger-like projections called villi.
- The villi are further covered by smaller projections called microvilli.

Other Uses of Digestive Enzymes

Microorganisms also make digestive enzymes. Decay organisms such as certain bacteria and fungi release these enzymes on to the food and take up the soluble products. These organisms are called **saprophytes**.

Scientists have used microorganisms such as saprophytes to supply enzymes for a number of uses.

- **Proteases** and **lipases** are used in biological washing powders.
- Proteases are used to pre-digest protein in some baby foods.
- **Amylases** are used to convert starch into sugar syrup.
- **Isomerase** is used to convert glucose into fructose which is sweeter.

? Test Yourself

1. Why does bread start to taste sweet if it is chewed for several minutes?
2. What is the function of the gall bladder?
3. What are the products of fat digestion?
4. Why is fructose used in sweets rather than glucose?

★ Stretch Yourself

1. The lining of the stomach is protected by mucus. Why does it need to be protected?
2. People who have coeliac disease may have many of their villi destroyed. What effect might this have on the process of absorption? Explain your answer.

Practice Questions

Complete these exam-style questions to test your understanding. Check your answers on page 124. You may wish to answer these questions on a separate piece of paper.

1 The following structures are involved in transport in plants. Match words **A**, **B**, **C**, and **D**, with numbers **1–4** in the sentences. (4)

A stomata **B** xylem **C** phloem **D** root hairs

Plants take water up from the soil. Plants have many _____**1**_____ to increase their surface area for water uptake. The water is carried up the stem in the _____**2**_____. Sugars, however, are transported in the _____**3**_____. Water is lost to the air through _____**4**_____.

2 The boxes contain some types of blood vessel and some descriptions. Draw straight lines to join each blood vessel to the correct description. (3)

Blood vessel
Aorta
Pulmonary artery
Pulmonary vein
Vena cava

Description
Carries oxygenated blood under low pressure.
Carries blood into the right atrium.
Carries deoxygenated blood away from the heart.
Carries oxygenated blood under high pressure.

3 Complete these sentences by writing the correct words in the gaps. (5)

Starch molecules are too large to be able to pass into the bloodstream and so need to be _____ first. This digestion begins in the _____. An enzyme called _____ breaks down starch into maltose. Maltose is then digested into _____ in the small intestine. Absorption then occurs and this is speeded up by the presence of tiny projections on the wall of the small intestine called _____.

4 The diagrams show three different blood vessels.

A _____ B _____ C _____

a) Write the name of each type of blood vessel under the correct diagram. (3)

b) Which blood vessel **A**, **B** or **C** has valves along its length? (1)

c) Describe how blood vessel **C** is adapted for its function. (2)

5 Harry studied water loss from a leafy shoot. He used an apparatus called a potometer. It is shown in the diagram.

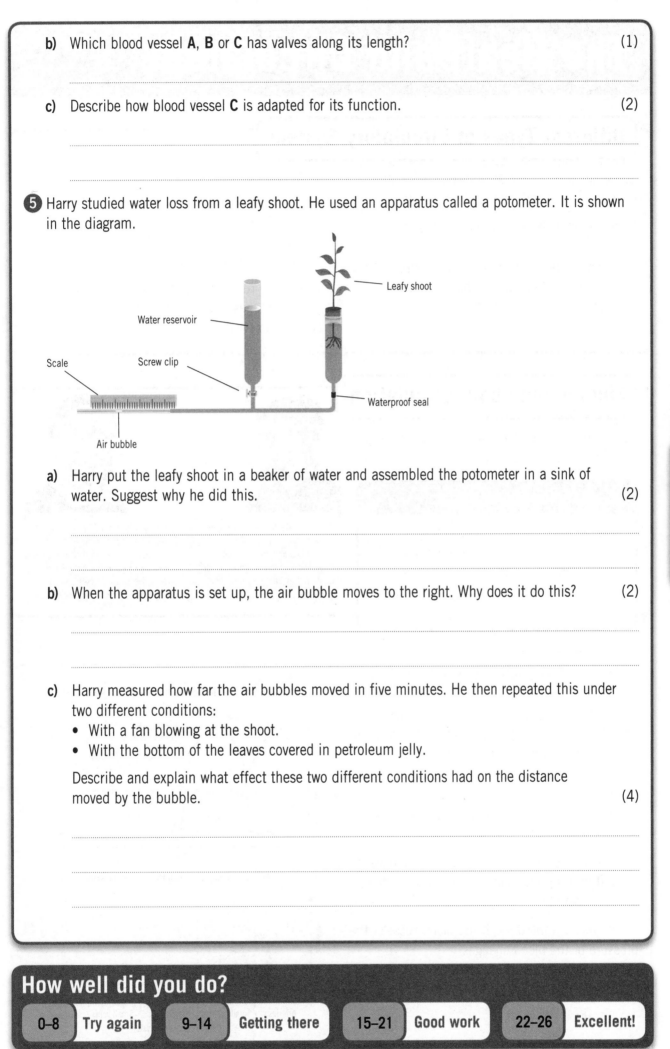

a) Harry put the leafy shoot in a beaker of water and assembled the potometer in a sink of water. Suggest why he did this. (2)

b) When the apparatus is set up, the air bubble moves to the right. Why does it do this? (2)

c) Harry measured how far the air bubbles moved in five minutes. He then repeated this under two different conditions:
- With a fan blowing at the shoot.
- With the bottom of the leaves covered in petroleum jelly.

Describe and explain what effect these two different conditions had on the distance moved by the bubble. (4)

How well did you do?

| 0–8 | Try again | 9–14 | Getting there | 15–21 | Good work | 22–26 | Excellent! |

The Heart and Circulation

Different Types of Circulatory System

Some organisms such as amoeba are small enough not to need a circulatory system.

Larger animals have different types of circulatory system:

- Insects have an open circulatory system. The blood is not circulated in blood vessels but moves around in the body cavity.

- Vertebrates have a closed circulatory system in which the blood is transported around the body in blood vessels. In fish this is a single system and the blood goes straight to the body from the gills. In humans there is a double system.

The structure of the heart and blood vessels and the advantages of a double circulatory system are covered on page 85.

Finding out about Circulation

Our circulation has been investigated by many scientists throughout history.

✓ Maximise Your Marks

How Science Works questions may ask for examples of how ideas in science have changed over the years. The discoveries made about circulation is a good example to use.

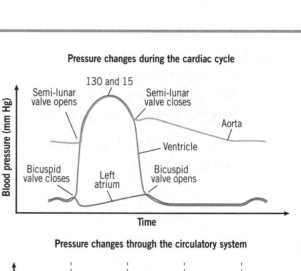

Galen was doctor to five Roman emperors. He carried out dissections and showed that arteries carried blood, not air. He could not explain how the blood circulated.

Harvey lived from 1578 to 1627. He carried out experiments and showed that the blood flows from the heart in arteries and back in veins. He could not see capillaries but guessed that they were there.

Build Your Understanding

The pattern of contraction of the different chambers of the heart is called the cardiac cycle:
- First the atria fill up with blood.
- They then contract and force the blood into the ventricles.
- The ventricles then contract and force blood out into the arteries.

The diagram shows how the pressure changes in the left atrium, left ventricle and aorta as these events happen.

As the blood flows through the blood vessels, the pressure of the blood changes.

Pressure changes during the cardiac cycle

130 and 15

Semi-lunar valve opens · Semi-lunar valve closes · Aorta · Ventricle · Bicuspid valve closes · Left atrium · Bicuspid valve opens

Blood pressure (mm Hg) / Time

Pressure changes through the circulatory system

Blood pressure / Part of circulatory system

Arteries · Small arteries · Capillaries · Small veins · Large veins

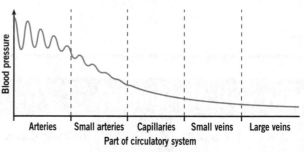

The Heart Beat and Pulse Rate

The heart is made of muscle which is supplied with glucose and oxygen by the coronary artery. This supply is needed for respiration to provide the energy for the contraction.

The pulse is a measure of the heart beat which puts the blood under pressure. It can be detected at various places, for example in the wrist, ear and temple.

Build Your Understanding

The rate of the heart is controlled by a small group of cells called the pacemaker. There are two areas called the sinoatrial node (SAN) and the atrioventricular node (AVN). They produce a small electric current which spreads through the heart muscle, making it contract.

The electric current that is produced by the pacemaker can be detected and studied using an ECG machine.

An ECG trace

✓ Maximise Your Marks

Remember, the heart contracts on its own, without nerve messages or hormones to make it contract, but nerves and hormones (e.g. adrenaline) can make the pacemaker speed up or slow down.

Blood Groups and Clotting

Different people have different types of blood groups which are controlled by their genes.

One of the main sets of blood groups is the **ABO system**. Another is **rhesus** positive or negative. This describes the chemical groups or antigens found on the red blood cells. It is determined by a single gene with three possible alleles: A, B or O. A and B are codominant (fully expressed) and O is recessive to both.

People with different blood groups have different antibodies in their blood:
- People with group A have anti-B antibodies, and those with group B have anti-A antibodies.

- People with group AB have neither anti-A nor anti-B antibodies.
- People of group O have both.

Blood clotting occurs when platelets are exposed to air, causing a series of chemical reactions leading to the formation of a mesh of fibrin fibres (clot). Clotting can be affected:
- Haemophilia is an inherited condition in which the blood does not easily clot.
- Substances such as vitamin K, alcohol, green vegetables and cranberries affect clotting.
- Drugs such as warfarin, heparin and aspirin are used to reduce clotting.

❓ Test Yourself

1. What type of circulation does a fish have?
2. Suggest why William Harvey could not show that capillaries existed.
3. What process in the heart muscle releases the energy for contraction?
4. Why are people sometimes given drugs such as heparin during and after an operation?

★ Stretch Yourself

1. Compare the flow of blood in the arteries with that in the veins.
2. A person is blood group A and rhesus negative. Which antibodies do they have in their blood?

The Skeleton and Exercise

The Organisation of the Skeleton

Different animals have different types of skeleton:
- Animals like insects have an **external** skeleton.
- All vertebrates have an **internal** skeleton. This makes it easier to grow and easier to attach muscles to.

The skeleton carries out important functions:

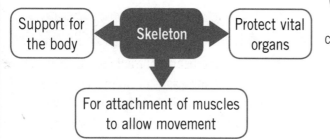

The skeleton of vertebrates is made up of **bone** and **cartilage**, both of which are living tissues. In animals like sharks, the skeleton is mainly cartilage, but in humans it is mainly bone.

The long bones of our arms and legs are hollow. This makes them much lighter, but still strong.

The structure of a long bone

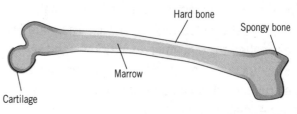

Hard bone
Spongy bone
Marrow
Cartilage

✓ Maximise Your Marks

Remember that our skeleton starts off as cartilage and turns to bone as our growing slows down. The amount of bone or ossification (the process of bone formation) can be used to tell how old a person is or was from their skeleton.

Joints

Where two bones meet they form a **joint**. Some of these joints are fused but others allow movement.

Synovial joints are specially adapted to allow smooth movement.

A synovial joint

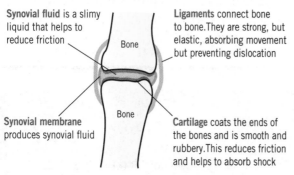

Synovial fluid is a slimy liquid that helps to reduce friction

Bone

Ligaments connect bone to bone. They are strong, but elastic, absorbing movement but preventing dislocation

Bone

Synovial membrane produces synovial fluid

Cartilage coats the ends of the bones and is smooth and rubbery. This reduces friction and helps to absorb shock

This joint is a **hinge joint**, such as the joints between the fingers and at the elbow or knee. These joints allow movement in one dimension.

Ball and socket joints, such as the hip joint, have almost all-round movement.

Muscles and Movement

Muscles are the main effectors in the body. They contain muscle fibres that can shorten and so make the muscle contract. In order to contract muscles need energy from respiration. They can only contract and pull on bones; they cannot actively expand or push.

Muscles therefore have to be arranged in pairs called **antagonistic pairs**. When one contracts, the other relaxes, and vice versa. Muscles are connected to bones by **tendons**. Tendons are very strong and are not elastic.

♀ Boost Your Memory

Remember that people who are antagonistic are always arguing and taking the opposite point of view. In the same way, antagonistic muscles work against each other.

Muscles and Levers

The biceps and triceps are antagonistic muscles in the human arm.

The arm works like a **lever**, with the elbow being the **pivot**.

The muscles are attached close to the pivot so this means that:

- A larger distance is moved by the hand than the muscles.
- A larger force is exerted by the muscles than is exerted by the hand.

When the biceps muscle contracts, the arm bends

Tendons are made of very strong protein and are inelastic. They join muscles to bones

When the triceps muscle contracts, the arm straightens

Build Your Understanding

Many people now visit a gym, and regular exercise can improve a person's fitness. It can also improve their general health, helping to prevent diseases such as heart disease. When a person decides to undertake a proper fitness programme they should follow these steps:

First discuss factors in the person's lifestyle, such as alcohol and tobacco consumption, family medical history and any medication they are taking.

→

Then devise a fitness programme, weighing up any side effects with any benefits that may be gained.

↓

Modify the programme if the fitness is improving faster than expected or in the case of an injury.

←

Monitor the progress of the training including changes in heart rate, blood pressure and recovery period.

↓

Decide how successful the treatment has been, taking into account the accuracy of the monitoring technique used and the repeatability of the data.

? Test Yourself

1. Why are synovial joints called synovial?
2. Why do the bones of birds need to be particularly light?
3. What type of joint is found at the hip?
4. What are the differences between ligaments and tendons?

★ Stretch Yourself

1. The triceps muscle is called an extensor muscle. Why is this?
2. What is the 'recovery period' that is measured when exercising?

The Excretory System

Use, Damage and Repair

Different Wastes

The body produces different types of waste as a result of its metabolism. Many of these wastes are toxic and so must be removed from the body. The removal of these wastes is called **excretion**.

✓ Maximise Your Marks

Remember that in excretion the waste product has to be made by the body. Most faeces is not made of excretory products. It has been taken in, passed through the gut and passed out. This is called **egestion**.

The sites of production of excretory products

The skin excretes sweat containing water and salts. This evaporates and so cools the body

The lungs remove carbon dioxide from the blood. An increased concentration of carbon dioxide is detected by the brain and the breathing rate is increased

The liver produces urea from excess amino acids

The kidneys excrete urea, water and salts

How the Kidneys Work

The kidneys produce urine by the following process:

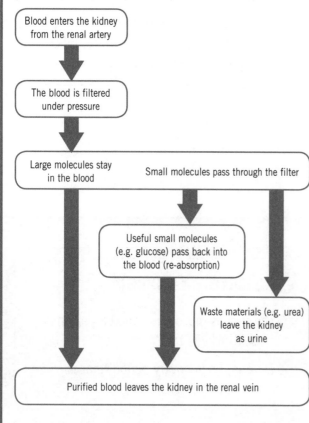

Blood enters the kidney from the renal artery

↓

The blood is filtered under pressure

↓

Large molecules stay in the blood

Small molecules pass through the filter

↓

Useful small molecules (e.g. glucose) pass back into the blood (re-absorption)

Waste materials (e.g. urea) leave the kidney as urine

Purified blood leaves the kidney in the renal vein

The useful substances that pass back into the blood are:
- All the sugar and amino acids (re-absorbed by active transport).
- The dissolved minerals needed by the body.
- As much water as is needed by the body.

✓ Maximise Your Marks

It is important to realise that although there is no glucose in the urine and also no protein, the reasons for this are different in each case.

The Structure of the Kidney

The kidneys are made of millions of small tubes called kidney tubules. Each individual kidney tubule is called a nephron.

Different parts of these tubules do different jobs in the production of urine.

The part of the tubule that filters the blood is made of a tight knot of blood capillaries called the glomerulus.

The capsule then collects the fluid that is forced out of the glomerulus.

The job of the kidneys in controlling the water balance of the body involves the hormone ADH. This is described on page 9.

ADH acts on the final part of the nephron, making it more permeable to water. More water is therefore re-absorbed and more concentrated urine is produced.

Structure of a kidney tubule

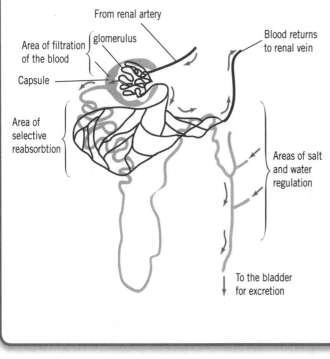

From renal artery

glomerulus

Area of filtration of the blood

Capsule

Blood returns to renal vein

Area of selective reabsorbtion

Areas of salt and water regulation

To the bladder for excretion

This is described on page 9.

✓ Maximise Your Marks

Remember that selective re-absorption means that some substances (like glucose and amino acids) are re-absorbed and others are not. As this is against a concentration gradient, selective re-absorption must use active transport.

Build Your Understanding

When the body takes in proteins, they are digested into amino acids and absorbed into the bloodstream.

The body cannot store amino acids so if there are too many then they are either built up into proteins or destroyed. The liver breaks down excess amino acids, releasing urea.

Urea is sent to the kidneys for excretion.

? Test Yourself

1. What is contained in sweat?
2. If the blood pressure drops, the kidneys stop working. Why is this?
3. Why is there usually no protein in the urine?
4. Why are there usually no amino acids in the urine?

★ Stretch Yourself

1. Alcohol reduces ADH production. What effects can this have?
2. Why is there more urea in a person's urine the day after eating a high protein meal?

Use, Damage and Repair

Breathing

Gaseous Exchange in Different Animals

Different types of organisms have different systems for obtaining oxygen and losing carbon dioxide.

The moving of these gases between the organism and the environment is called **gaseous exchange**. In small organisms, such as worms, gaseous exchange occurs over the whole body surface.

Fish have **gills** for gaseous exchange. The filaments take in oxygen from the water. The presence of gills means that fish must live in water. Amphibians such as frogs have lungs, but they also get oxygen through their skin. This needs to be moist so they can live on land but only in damp places.

Humans have **lungs** for gaseous exchange. The gaseous exchange takes place in millions of tiny air sacs called **alveoli**.

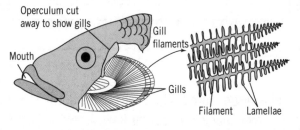

Gill structure

Operculum cut away to show gills

Mouth

Gill filaments

Gills

Filament   Lamellae

Build Your Understanding

The human alveoli are adapted for efficient gaseous exchange in a number of ways:
* They are permeable.
* They have a moist surface.
* There are millions of alveoli, providing a surface area of about 90 m².
* There are many blood vessels, providing a rich blood supply.

Fish gills allow efficient gaseous exchange in water because the filaments:
* have a large surface area
* are very thin
* are well supplied with blood.

Breathing Mechanisms

Large and active organisms cannot simply rely on diffusion to bring enough oxygen to their respiratory surfaces.

Organisms like mammals and fish **ventilate** their respiratory surfaces using muscular contractions. This is called **breathing**.

In humans, drawing air in and out of the lungs involves changes in pressure and volume in the chest. These changes are brought about by contractions of the diaphragm and intercostal muscles. They work because the **pleural** membranes form an airtight **pleural cavity**.

In breathing in (inhaling):
* The intercostal muscles contract, moving the ribs upwards and outwards.
* The diaphragm contracts and flattens.
* Both of these actions increase the volume in the pleural cavity and so decrease the pressure.
* Air is drawn into the lungs because the air moves from the higher atmospheric pressure into the lungs, where there is lower pressure.

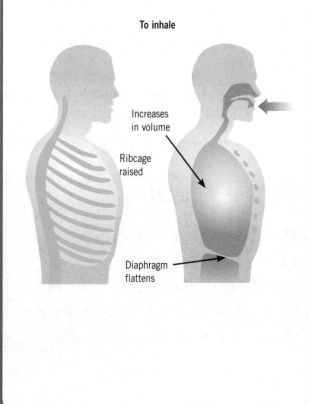

To inhale

Increases in volume

Ribcage raised

Diaphragm flattens

Breathing Mechanisms (cont.)

In breathing out (exhaling):
- The intercostal muscles relax and the ribs move downwards and inwards.
- The diaphragm relaxes and domes upwards.
- The volume in the pleural cavity is decreased so the pressure is increased.
- Air is forced out of the lungs.

In fish, water enters the mouth. The mouth then closes and forces water out across the gills.

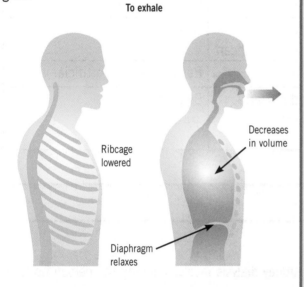

To exhale

Ribcage lowered

Decreases in volume

Diaphragm relaxes

✓ Maximise Your Marks

Make sure that you know the difference between breathing and respiration. Many candidates confuse the two. Breathing is the muscular movements that aid gaseous exchange, but respiration is the reaction that releases energy from food.

Build Your Understanding

The volume of air that is drawn in and out can be measured on a machine called a spirometer.
- Total lung capacity is the maximum volume of air that the lungs can hold.
- Vital capacity is the maximum volume of air that the lungs can pass in and out per breath.
- Residual volume is the air that is left in the lungs after a person has breathed out deeply.

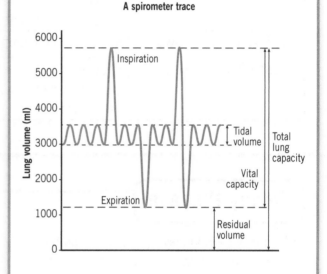

A spirometer trace

✓ Maximise Your Marks

Looking at spirometer traces like this can often give indications of disorders such as asthma (see page 101).

❓ Test Yourself

1. How does a worm obtain oxygen?
2. Why do so many frogs live in tropical rainforests?
3. Where in a fish's gills does oxygen enter the bloodstream?
4. What happens to the diaphragm to make us breathe in?

⭐ Stretch Yourself

1. Why do gill filaments look pink?
2. What is the vital capacity of the person whose breathing is shown on the spirometer trace?

Damage and Repair

Heart Problems

There are many heart conditions and diseases that might affect the working of the heart. Examples are shown in the table:

If the damage is bad enough, the person may need a heart transplant.

Condition	Possible effect on the body	Possible treatment
An irregular heart beat	Less oxygen in the blood	Drug treatment
A hole in the heart	Blood can move directly from the right side to the left side of the heart	Surgery
Damaged or weak valves	Reduced blood circulation	Replacement by artificial valves
Coronary heart disease	Reduced blood flow to the heart muscle possibly leading to heart attacks	By-pass surgery

Kidney Disease

People may have kidney failure for a number of reasons. A person can survive if half of their kidney tubules are still working, but if the situation worsens there are two options:

Kidney dialysis → Kidney failure ← Kidney transplant

Kidney dialysis involves linking the person up to a dialysis machine.

This takes over the job of the kidneys and removes waste substances from the blood.

Build Your Understanding

In kidney dialysis, the blood is removed from a vein in the arm. It then passes through a long coiled tube made of partially permeable cellophane. The fluid surrounding the tube contains water, salts, glucose and amino acids, but it has no waste materials, such as urea. These waste materials therefore diffuse out of the blood into the fluid.

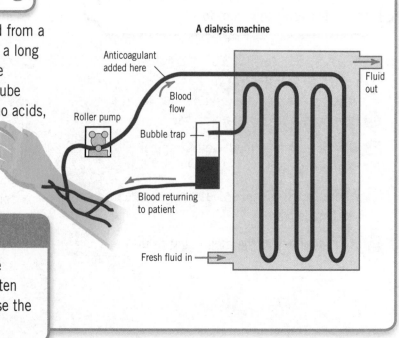

A dialysis machine

Anticoagulant added here

Roller pump

Blood flow

Bubble trap

Blood returning to patient

Fresh fluid in

Fluid out

✓ Maximise Your Marks

Remember that the dialysis machine works using diffusion. Candidates often think that it involves osmosis because the cellophane is partially permeable.

Fractures and Sprains

Despite being strong, bones can still be broken by a sharp knock. This is more likely to happen to elderly people because they often have **osteoporosis** which makes their bones weaker. There are three types of bone fracture:

- **A simple** fracture of the bone is a fracture without the skin being broken.
- **A compound** fracture of the bone is where the skin has been broken.
- **A green stick** is where the bone has not been fractured all the way through.

If a person has a fracture, they should not be moved as this could cause further damage (especially if it is a spinal injury).

Excessive exercise can cause damage to other parts of the body. These include:

- sprains
- dislocations
- torn ligaments
- damaged tendons.

These injuries may be treated by a physiotherapist who will devise a suitable set of exercises that should help the person recover.

⚓ Boost Your Memory

The best way to remember the treatment for sprains is to remember **RICE**:
Rest
Ice
Compression
Elevation

Lung Problems

The lungs are easily infected because they are a 'dead end'. Microbes and particles can easily collect there. Some of these types of problems are:

- **Industrial diseases**, such as asbestosis, where small particles of asbestos are breathed in and damage the lungs.
- **Genetic conditions**, such as cystic fibrosis, where too much mucus is made.
- **Lifestyle factors**, such as smoking, which can lead to lung cancer. Some of the actions of smoking on the lungs are described on page 28.

The respiratory system tries to protect itself from disease by producing mucus and by the action of cilia. These are shown on page 21.

Build Your Understanding

More and more people now have asthma. The symptoms of asthma are difficulty breathing, wheezing and a tight chest. It is treated using inhalers.

During an asthma attack:

- The lining of the airways becomes inflamed.
- Fluid builds up in the airways.
- The muscles around the bronchioles contract, constricting the airways.

❓ Test Yourself

1. Why is it important to keep the blood in the two sides of the heart separate?

2. Write down one substance that a dialysis machine needs to remove from the blood.

3. Why are old people more likely to suffer from a fractured bone?

4. How do people get cystic fibrosis?

⭐ Stretch Yourself

1. Why does dialysis fluid need to contain water, salts, glucose and amino acids?

2. Asthma inhalers have an effect on the muscles of the bronchioles. Suggest what effect they have.

Transcripts and Donations

Transplants

People are living longer now than they ever have done. This is due to less industrial disease, a healthier diet and lifestyle, modern treatments and cures for disease, and better housing.

This means that there is an increasing demand for transplants of different body parts:

- Kidney – as a person can survive with one kidney, it is possible for a person to donate one kidney to be transplanted into another person. Other transplants may come from dead donors.
- Heart – sometimes the heart is too badly damaged and surgery cannot repair the problem.
- Lung – the heart is often transplanted along with the lungs in a heart-lung transplant.
- Joints – operations to replace or resurface the ends of bones are now common.

The main problem with transplants is preventing the person's immune system rejecting the transplanted organ.

This is avoided by taking certain precautions:

- Making sure that the donor has a similar 'tissue type' to the patient.
- Treating the patient with drugs to make their immune system less effective.

Some replacements are mechanical rather than biological. The ends of bones may be metal, and heart-assist pumps may be inserted. Some of the problems of using mechanical replacements include their size, using materials that will not react with the body and, in some cases, the need for a power supply.

The Ethics of Transplants

The increasing demand for transplants has led to a major problem – a serious shortage of donors.

For some people this means waiting for a long time and some may die before a suitable organ becomes available.

The line on the graph shows the number of people in the UK on the transplant list in the period 1999–2008. The bars show the number of people who died and became donors and the number of transplants performed.

The current system only allows organs to be removed if the person who has died carried a donor card.

Relatives of the person can stop the organs being taken even if the person carried a card. A new system has been suggested which would mean that organs can be taken from anyone who dies unless that person had opted not to allow it. This is called **presumed consent**.

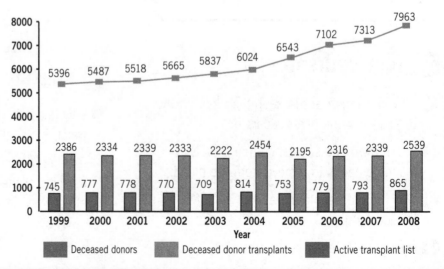

Number of deceased donors and transplants in the UK, 1999–2008 and patients on the transplant lists at December 31 2008

Presumed consent would greatly increase the numbers of organs available for transplants, but not everyone is happy with the idea. Some people have religious and cultural objections. Others might not realise that they need to opt out and could have their organs used against their will.

Mistakes could also be made and people could have their organs taken by mistake even if they have opted out. Getting it wrong could lead to distress for relatives and could lead to a backlash against doctors and organ donation.

✓ Maximise Your Marks

How Science Works questions often ask for arguments for and against new scientific developments. Make sure that you can give both sides of this argument.

Build Your Understanding

When giving blood transfusions, it is important to make sure that the blood groups are matched. Blood groups are discussed on page 93.

Unsuccessful blood transfusions cause agglutination (blood clumping). The antigens on red blood cells and the antibodies in the blood serum will decide how blood groups react and therefore if the blood transfusion is successful.

The table shows which blood groups are compatible.

	Type	O−	O+	B−	B+	A−	A+	AB−	AB+
	AB+	■	■	■	■	■	■	■	■
	AB−	■		■		■		■	
	A+	■	■			■	■		
	A−	■				■			
	B+	■	■	■	■				
	B−	■		■					
	O+	■	■						
	O−	■							

Blood group of donor (column header spanning Type row)

Blood group of receiver (row header, left side)

■ = successful transfusion

Gamete Donation

Some types of infertility can be treated with hormones. This is covered on page 11.

In other cases, sperm or egg production is not possible. These cases can be treated by:
- **Egg donation**, which would involve IVF or ovary transplants.
- **Artificial insemination**, which involves sperm donation. Artificial insemination not always involves donated sperm – often the sperm is from the biological father.

Sometimes the woman's body cannot support a baby throughout pregnancy and another female may have a baby for her. This is called **surrogacy**.

? Test Yourself

1. Why are kidney transplants between relatives more successful?
2. Write down one problem with having a battery-powered heart-assist pump inserted into the body.
3. Write down one property needed in the metal that is used to replace bone in joints.
4. Suggest one possible problem with surrogacy arrangements.

★ Stretch Yourself

1. A person who is blood group A− cannot successfully donate blood to a B− person. Explain why.
2. People with blood group O− are usually described as universal donors. Why is this?

Use, Damage and Repair

Practice Questions

 Complete these exam-style questions to test your understanding. Check your answers on page 125. You may wish to answer these questions on a separate piece of paper.

Check your answers on page 125.

1 The following terms are involved in human physiology. Match words **A**, **B**, **C**, and **D**, with numbers **1–4** in the sentences. (4)

A surrogacy   **B** dialysis   **C** bypass   **D** osteoporosis

A condition that weakens the bones is called _____1_____. Coronary heart disease may be treated by a _____2_____. The function of the kidneys may be replaced by _____3_____. A woman who cannot maintain a pregnancy may get a baby using _____4_____.

2 Put a tick (✓) or a cross (✗) in each empty box to show how the kidney deals with glucose, protein and urea. (4)

	Glucose	Protein	Urea
Present in the blood reaching the kidney			
Passes out of the blood in the filter unit			
Re-absorbed back into the blood from the kidney tubule			
Usually present in the urine			

3 The boxes contain names of parts of the body and some descriptions. Draw straight lines to join each part to the correct description.

Part
Bone
Cartilage
Ligaments
Tendons

Description
Living tissue containing cells and calcium salts.
Elastic structure that holds joints together.
Inelastic structure that joins muscle to bone.
Shiny substance that reduces friction in joints.

4 Look at the figure on page 92 showing pressure changes during the cardiac cycle and use the information in the figure to answer the following questions.

a) Explain why there is such a difference between the two pressures in the left ventricle and left atrium. (1)

b) How do you think the pressure would differ in the right ventricle compared with the left? Explain your answer. (2)

Use, Damage and Repair

5 Jack has asthma. This means that something causes the small tubes in his lungs to narrow. This makes it harder for him to draw air into the small air sacs where gaseous exchange occurs.

a) What is the name of these air sacs? (1)

....

b) Suggest one factor that might make the small tubes narrow. (1)

....

c) Jack decides to measure his breathing using a spirometer. He makes a trace when he is breathing normally and when he has an asthma attack. What do the letters VC and TV stand for? (2)

....

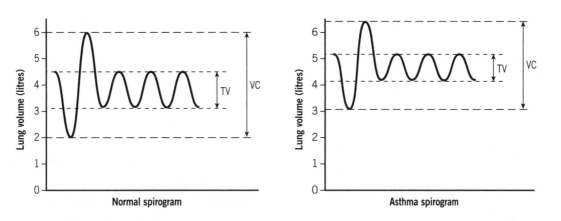

Normal spirogram

Asthma spirogram

d) Describe the differences in Jack's breathing when he has an asthma attack compared with normal breathing? (4)

....

....

....

6 Look at the graph on page 102. Write about what the graph shows. In your answer:

- Describe any changes in the rate of transplants compared with numbers on the list.

- Explain any difference between the number of donors and the number of transplants. (4)

....

....

....

....

Use, Damage and Repair

How well did you do?

| 0–8 | Try again | 9–13 | Getting there | 14–18 | Good work | 19–23 | Excellent! |

Types of Behaviour

Innate or Learned Behaviour

When animals are born they have certain inbuilt types of behaviour. This is called **innate behaviour**. It includes simple reflexes and **instincts**. Instincts are much more complicated actions than reflexes.

Innate behaviour is controlled by the genes and is inherited from the animal's parents. It is important as it gives the young animal certain skills it needs to survive before it has had the chance to learn.

The scientist Nikolaas Tinbergen studied seagulls to see what innate behaviour the gulls used to help their young to collect food from their parents.

As soon as a young animal is born it starts to change its behaviour. This is because it starts to learn. **Learning** is a change in behaviour caused by experiences.

There are several types of learned behaviour:
- Probably the first learning experience of animals is **imprinting**. Newborn animals will be attracted to the first animal or object that they see. The first scientific studies of this were carried out by Konrad Lorenz. He discovered that if geese were reared by him from hatching, they would treat him like a parent bird and follow him around.
- Another way that animals learn is by **habituation**. An animal may be frightened by a particular stimulus such as a loud noise. However, if the stimulus is repeated and the animal is not harmed then the animal learns to ignore the stimulus. It has been habituated. This idea is often used in the training of animals such as police horses.
- Animals can also learn by **conditioning**. Instead of learning to ignore a stimulus, they will associate one stimulus with another. This was first discovered by a Russian called Ivan Pavlov. This type of learning is also used in training animals for specific jobs.
- **Insight** learning can be shown by apes when they use their intelligence to solve a problem.

Build Your Understanding

There are two main types of conditioning:
- **Classical conditioning** is the type shown by Pavlov's dogs. Pavlov began to ring a bell each time the dogs were shown their food. After a while Pavlov found that the dogs salivated when the bell was rung regardless of whether food was present. The dogs had become conditioned so that they associated a bell with the arrival of food. The response is caused by a stimulus different from the one that originally triggered it.
- **Operant conditioning** is sometimes called 'trial and reward learning'. It might involve giving animals (for example rats) a food reward if they press a lever after seeing a light or hearing a sound. Some sort of punishment can also be used as a negative reinforcement.

Unlike classical conditioning the response is not natural behaviour. Dogs producing saliva is natural, but rats pushing levers is not.

Operant conditioning

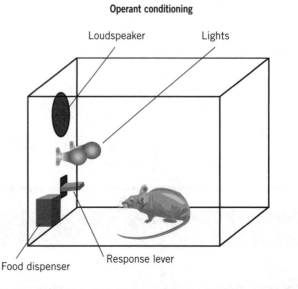

Loudspeaker — Lights — Food dispenser — Response lever

Memory and the Brain

In order to learn, organisms must be able to remember past events.

Memory is the storing of information in the brain so that it can be retrieved.

This takes place in a part of the brain called the **cerebral cortex**. This part of the brain is also responsible for intelligence, language and consciousness.

When we learn something, certain pathways in the brain are formed and others are lost. The more we repeat the task, the more likely that the neural pathway will stay connected.

It seems that some parts of the brain may only be able to learn skills up to a certain age. For example, children who have been brought up away from other people have difficulty learning language.

Build Your Understanding

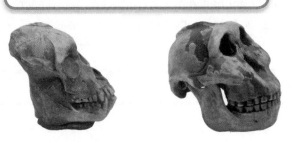 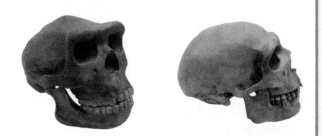

The development of the human brain in evolution has provided humans with a greater ability to learn. This has enabled a better chance of survival.

The evidence for human evolution has come from a number of sources:
- The discovery of ancient **fossils** has provided much evidence, especially fossils of skulls. A scientist called Louis Leakey discovered human fossils in Africa dating from about 1.6 million years ago. A 3.2-million-year-old fossil nicknamed Lucy was found and recently a 4.4-million-year-old fossil nicknamed Ardi.

- The discovery of ancient tools has told us a lot about ancient human behaviour and skills.
- The study of a special type of DNA, which is found in mitochondria, has suggested that humans evolved in Africa.

✓ Maximise Your Marks

Remember that mitochondrial DNA is not inherited in the normal way, i.e. half from our mother and half from our father. Our mitochondrial DNA comes entirely from our mother.

❓ Test Yourself

1. Why is imprinting important for young animals?
2. Why are police horses exposed to loud noises during their training?
3. Why do scarecrows stop having an effect on birds after a while?
4. Which part of our brain contains our memories?

⭐ Stretch Yourself

1. How could operant conditioning be used to train a dog a trick?
2. Why can studying fossil skulls give us information on the intelligence of ancient humans?

Communication and Mating

Finding a Mate

Sexual reproduction involves individuals choosing who to mate with. This is particularly important for females as they use energy and food reserves when producing their young. **Courtship** is a type of behaviour used to help choose a mate.

In many types of animals, males show courtship behaviour to try to persuade females that they have good genes and can provide for the offspring. This is why in many species the male is more brightly coloured than the female.

Once a mate is chosen there are a number of different types of relationship:
- Very few animals mate with the same partner for life although the albatross is an exception. Albatrosses can live to be 80–85 years old and they mate for life.
- Some animals live in groups with a dominant male that mates with all the females.
- Other animals may have different mates each year or even in the same breeding season.

Parental Care

Most birds and mammals look after their young for some time after they are born. This is called **parental care**.

This means that these animals have developed special feeding behaviour.

Baby birds will call to their parents and often gape with their mouth wide open. Often the inside of the baby's mouth is coloured to attract the parent bird.

Mammals feed their young on breast milk. The young are born with an instinct to suck on the breast. The process of breast feeding is also important in building an emotional bond between the mother and the baby.

Showing parental care to their babies has advantages and disadvantages for the parents:
- Looking after the offspring increases the offspring's chance of surviving and passing on the parents' genes.
- However, looking after the offspring takes energy and makes the parents more at risk from predators.

Communication

Many animals live together in various types of social group. This makes it necessary for them to communicate with each other. There are a number of ways that animals can do this:

- Sounds.
- Visual signals.
- Airborne chemicals such as pheromones.

In mammals, such as chimpanzees, gorillas and humans, a lot of information is exchanged by **body language**. This includes gestures, body posture and facial expressions.

The diagram shows some facial expressions in chimps, but the same expressions may mean something completely different to another type of animal. They are **species-specific**.

✓ Maximise Your Marks

Do not assume that animal behaviour means the same thing as human behaviour. That is called **anthropomorphism**. Chimps may look like they are smiling, but it is probably a sign of aggression.

Build Your Understanding

Humans are great apes, along with gorillas, orangutans, chimpanzees and bonobos (pigmy chimps).

Until recently little was known about the behaviour of great apes in the wild. People did not understand how they communicate and did not realise how intelligent they are.

Two scientists studied great apes in the wild:

- Jane Goodall studied chimpanzees. She made two major discoveries. Firstly, like humans, chimpanzees enjoyed a mixed diet that included meat. Secondly, they could make simple tools, stripping leaves off branches to reach into termite nests. Until this time, only humans were thought to be clever enough to make tools.
- Dian Fossey studied mountain gorillas. She worked out many aspects of how they communicated and helped to protect them from poachers.

✓ Maximise Your Marks

Do not mix up chimpanzees with monkeys – great apes are not monkeys.

Apes are larger, spend more time upright, depend more on their eyes than on their noses, and do not have tails. They are more intelligent than monkeys.

? Test Yourself

1. Why does a male peacock have such brightly coloured tail feathers?
2. What is unusual about albatross mating?
3. Write down one disadvantage of showing parental care.
4. What are pheromones?

★ Stretch Yourself

1. What similarities between our hands and those of other great apes make it easier to use tools?
2. Dian Fossey is described as the first person to 'habituate herself with gorillas'. What does this mean?

Practice Questions

Complete these exam-style questions to test your understanding. Check your answers on page 126. You may wish to answer these questions on a separate piece of paper.

1. The following words are used in describing behaviour in apes. Match words **A**, **B**, and **C** with numbers **1–3** in the sentences. (3)

   A pheromones   **B** learning   **C** body language

   Apes have the ability to communicate using expressions. This is an example of _____1_____ .

   They can also communicate using chemicals called _____2_____. It has been shown that apes can change their behaviour as a result of experiences. This is called _____3_____.

2. The boxes contain some types of behaviour and some examples. Draw straight lines to join each type of behaviour to the correct example. (3)

Type of behaviour
Courtship
Habituation
Imprinting
Conditioning

Example
Police dogs have been trained not to be frightened of loud noises.
Male peacocks often display their tail feathers.
Car drivers usually brake when they see a red traffic light without thinking about it.
Baby chicks will follow the first moving object that they see.

3. Grebes are birds that show courtship behaviour. This is shown in the diagrams.

   Ear tufts

   Neck frill

   Before courtship   **Courtship begins**   **The grebes then dance and give each other waterweed**

   a) Why is courtship behaviour important to birds? (2)

   ....

   ....

b) Write down one difference between the grebes once they have started their courtship behaviour compared with before. (1)

....

....

c) Suggest why the grebes give each other waterweed during their dance. (2)

....

....

4 The figure shows an apparatus used by the scientist Ivan Pavlov in 1905. He operated on the dog to insert a tube into its salivary gland to collect saliva. Pavlov gave the dog meat powder just after a ticking noise. The dog produced saliva.

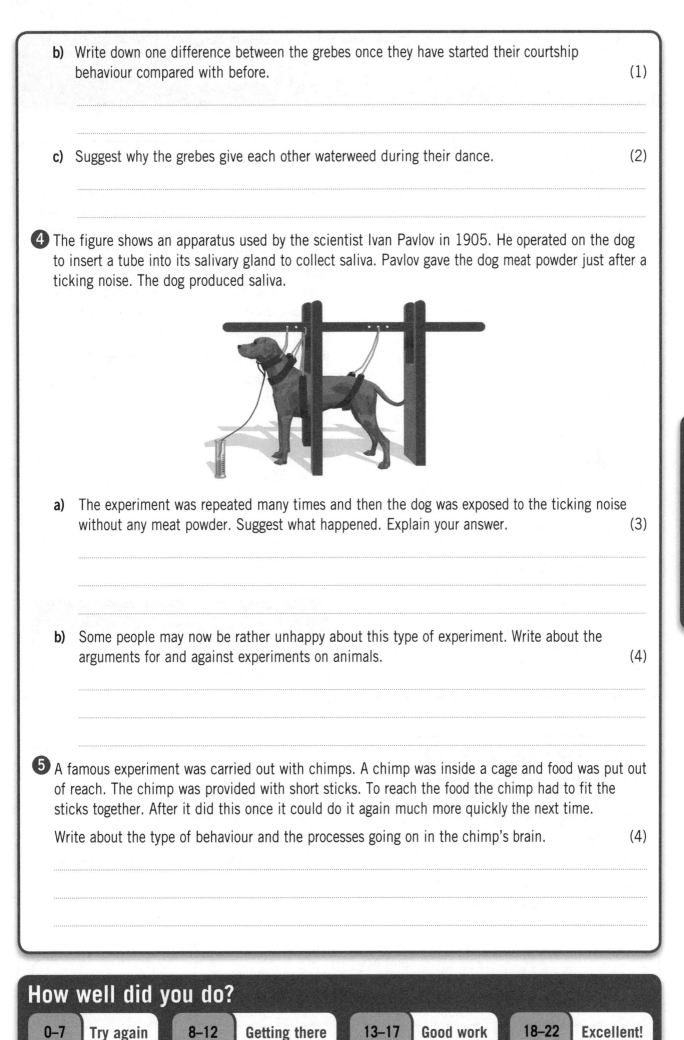

a) The experiment was repeated many times and then the dog was exposed to the ticking noise without any meat powder. Suggest what happened. Explain your answer. (3)

....

....

....

b) Some people may now be rather unhappy about this type of experiment. Write about the arguments for and against experiments on animals. (4)

....

....

....

5 A famous experiment was carried out with chimps. A chimp was inside a cage and food was put out of reach. The chimp was provided with short sticks. To reach the food the chimp had to fit the sticks together. After it did this once it could do it again much more quickly the next time.

Write about the type of behaviour and the processes going on in the chimp's brain. (4)

....

....

....

How well did you do?

| 0–7 | Try again | 8–12 | Getting there | 13–17 | Good work | 18–22 | Excellent! |

The Variety Of Microbes

Types of Microbe

Microorganisms are organisms that are too small to be seen by the naked eye. There is quite a variety of organisms that fall into this category:
- Fungi, such as yeast.
- All bacteria.
- All viruses.
- Single celled members of the protoctista kingdom.

Fungi have certain features in common, but **yeast** is unusual because it is made up of single cells. Each cell has a nucleus, cytoplasm and cell wall. They reproduce asexually by growing a small **bud** on the side which breaks off and forms a new cell.

Bacteria are just a few microns (thousandths of a mm) in size. Their structure is detailed on page 61. They have many different shapes such as spherical, rod, spiral and curved rods. Some bacteria can move using a **flagellum**. Bacteria all reproduce by a type of asexual reproduction called **binary fission**. They can survive on an enormous range of energy sources and can exploit a wide range of habitats as some take in food while others can photosynthesise.

Viruses are much smaller than bacteria and fungi, and are usually considered to be not-living. They are made of a protein coat surrounding a strand of genetic material. Viruses can only reproduce in other living cells by injecting their genetic material into the cell and using the cell to make the components of new viruses which then assemble into new virus particles.

Protoctista include organisms that used to be called protozoa and also single-celled algae such as plankton.

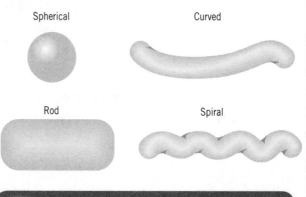

Spherical   Curved

Rod   Spiral

💡 Boost Your Memory

Think of a way of remembering the four types of microbes. You could use the first letters of their names: **F, B, V** and **P.**

Microbes and Disease

All the types of microbe include some organisms that are pathogens and cause disease. The ways that these microbes can be passed on is shown on page 21.

Examples of pathogens are:
- Bacteria such as *Salmonella* and *E. coli* that cause food poisoning.

- *Vibrio* bacteria that cause cholera.
- Fungi like *Trichophyton* that cause athlete's foot.
- Viruses that cause influenza and chickenpox.

For a pathogen to cause disease, various steps usually happen (see below):

Pathogen enters the body → Pathogen grows rapidly – the incubation period → Some pathogens make poisonous toxins → Disease symptoms appear

Microbes and Disease (cont.)

Diseases such as cholera can occur at any time, but after a natural disaster large numbers of people may become ill. This is because sewage and water systems may become contaminated and refrigerators may not work due to the lack of electricity.

Modern medicine can treat many diseases. This is the result of discoveries made by many scientists. **Pasteur**, **Lister** and **Fleming** are three of these:

- Louis Pasteur studied a number of diseases such as rabies and anthrax. He was the first person to realise that diseases that can be passed on are caused by living organisms. This is called the **germ theory**.

- Joseph Lister worked as a doctor, operating on patients. He found that treating his instruments and washing his hands with a chemical called carbolic acid helped to stop his patients' wounds becoming infected. This was the first **antiseptic** to be used.

- Sir Alexander Fleming was growing bacteria on petri dishes. He noticed that one of his dishes had a fungus growing on it. Around the fungus, called *Penicillium*, the bacteria had been killed. The fungus was producing a chemical called penicillin which was the first **antibiotic** to be discovered.

Build Your Understanding

Plankton are microscopic plants (phytoplankton) and microscopic animals (zooplankton).

Phytoplankton are capable of photosynthesis and are the main producers in ocean food chains and webs. The number of plankton living in lakes and the sea varies at different times of the year. These seasonal fluctuations can be explained by changes in light, temperature or minerals.

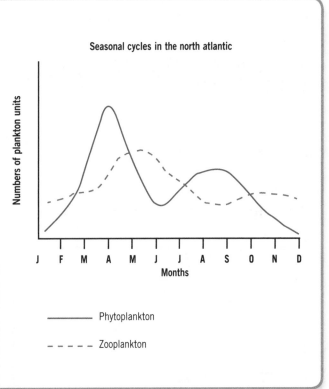

Seasonal cycles in the north atlantic

——— Phytoplankton

- - - - - Zooplankton

✓ Maximise Your Marks

Be careful interpreting graphs such as this one. This is the North Atlantic so think about which months will be warmer and lighter. Graphs of more southern regions would be different.

? Test Yourself

1. Why is reproduction in yeast often called 'budding'?
2. What is a flagellum?
3. Put a virus, a bacterium and a yeast cell in order of size, with the largest first.
4. How did the antibiotic penicillin get its name?

★ Stretch Yourself

1. Look at the graph of plankton numbers. Suggest why the numbers of phytoplankton peak in April.
2. The numbers of zooplankton peak in May–June. Suggest why this is.

Putting Microbes to Use

Biotechnology

Microorganisms have been used for hundreds of years for making foods such as bread and cheese. They are now being used more and more to produce new types of food and other useful products. The use of living organisms to make useful products is called **biotechnology**.

Microorganisms can be grown in large vessels called **fermenters**. The conditions are carefully controlled so that the microorganisms grow very fast. Like all organisms, they need food to obtain energy, but they can often be fed on waste from other processes.

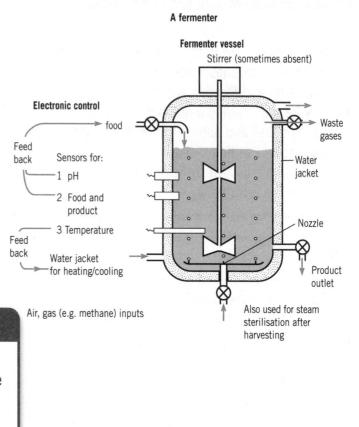

A fermenter

💡 Boost Your Memory

To help you remember how different products are made by microbes try and use the headings energy source, microbe, respiration and products. These are given for each example.

Making Yogurt

Energy source = milk

Microbe = bacteria (*Streptococcus* and *Lactobacillus*)

Respiration = anaerobic

Product = lactic acid

To make yogurt, the equipment is first sterilised and the milk is **pasteurised**. A culture of bacteria is added and the mixture is incubated at about 46°C for 4–5 hours. Before packaging, fruit, flavours or colours may be added.

✔ Maximise Your Marks

The lactic acid made by the bacteria makes the milk proteins clot to make the yogurt thicker and gives it an acidic taste.

Making Mycoprotein

Energy source = any starch or sugar source such as potato waste

Microbe = a fungus (*Fusarium*)

Respiration = aerobic

Product = mycoprotein

Mycoprotein is often used as a meat substitute as it has a high protein content. It does have a number of advantages over meat:
• The fungus grows very quickly.
• It has a high fibre content.
• It is low in fats.
• It can be grown on waste substances.

Making Alcohol

Energy source = grapes for wine or barley for beer

Microbe = yeast

Respiration = anaerobic

Products = alcohol and carbon dioxide

Making alcohol relies on yeast fermenting sugar. This sugar can come from different sources. The equation for fermentation is:

Glucose → **Ethanol (alcohol)** + **Carbon dioxide**

$$C_6H_{12}O_6 \rightarrow \quad 2C_2H_5OH \quad + \quad 2CO_2$$

In beer-making, barley is allowed to germinate so that the starch is turned into sugar. Hops are added to give the beer flavour. After the yeast has fermented, the beer is drawn off the yeast. It may then be pasteurised to kill any microbes before it is bottled.

The concentration of alcohol made by fermentation is limited. This is because when the alcohol reaches a certain concentration it will kill the yeast.

Spirits are made by **distillation**. This concentrates the alcohol.

Build Your Understanding

Fossil fuels such as coal and oil have been produced from living material over millions of years. They are not sustainable as we are burning them faster than they are being produced. This means that carbon dioxide is being added to the air. Biofuels are sustainable because they release carbon dioxide to the air at the same rate that it is absorbed by photosynthesis.

However, people are worried that in some areas tropical forests might be being destroyed to grow biofuel crops.

Making Biofuels

Biofuels are fuels that have been made from living material that has been produced in a sustainable way.

Biogas is produced in fermenters called **digesters**. Waste materials such as sewage or plant products are put in the tank. A mixture of different bacteria use these substances for anaerobic respiration. They produce biogas, which is made up of:

- Largely methane.
- Some carbon dioxide.
- Small amounts of hydrogen, nitrogen and hydrogen sulphide.

The biogas can then be burnt to make electricity, produce hot water or power motor vehicles.

Some countries can grow large amounts of sugar cane or maize. These crops can be used to produce sugar that can be fermented by yeast. This produces **ethanol**. The ethanol can be concentrated by distillation and added to petrol.

Petrol (about 85%) + Ethanol (about 15%)
Gasohol

❓ Test Yourself

1. When making yogurt, the equipment is sterilised first. Why is this?

2. After pasteurisation when making yogurt, the milk is cooled to 46°C before the culture is added. Why is this?

3. Why can vegetarians use mycoprotein as a meat substitute?

4. Fermentation can only make wine containing up to about 15% alcohol. Why is this?

⭐ Stretch Yourself

1. 'Energy forests' grow trees that are regularly harvested and burnt as fuel. How can this be made sustainable?

2. Suggest why people are so worried about the destruction of tropical forests.

Microbes and Genetic Engineering

Uses of Genetic Engineering

It is now possible to make microbes produce different proteins by changing their DNA. They are then called **genetically modified (GM)**.

If the gene has come from a different species then the GM organism is called **transgenic**.

The main steps in making a GM organism are:
- Finding and removing the necessary gene.
- Putting the gene into a vector that will carry it into the new cell.
- Testing the cells to see if they have taken up the gene.
- Allowing the new cells to reproduce and produce the new protein.

Some of the arguments for and against genetic engineering are on page 39.

GM organisms can have lots of possible uses:

GM Microbes	GM Plants	GM Animals	GM Humans
Making chymosin	Resistance to weedkillers	Faster growth	Gene therapy
Making insulin	Increasing yield	Producing useful proteins in milk	
Making human growth hormone	Producing insecticide		
	Resistance to disease		

Build Your Understanding

A number of steps in genetic engineering involve the use of enzymes.

An example of this is the modification of rice so that it produces vitamin A. Two different enzymes are used.
- **Restriction enzymes** cut the DNA in specific places.
- **DNA ligase** joins DNA together.

The restriction enzymes do not make a straight cut in the DNA but make a staggered cut. This produces exposed unpaired bases.

They are called **sticky ends** because any piece of DNA with complementary unpaired bases will stick to it.

How to produce genetically engineered rice

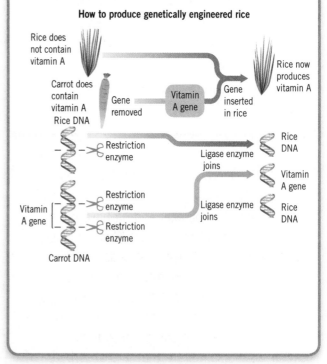

Insulin and Chymosin Production

Insulin is needed to treat people who have diabetes. This is explained on page 7. Until 1982 the insulin used was extracted from cows or pigs. This was difficult to extract, worked more slowly than human insulin and the body often rejected it.

Today, the human gene for insulin is placed into bacteria which produce insulin that is identical to human insulin. This works faster, can be produced in large quantities and is not rejected as it is identical to human insulin.

Insulin and Chymosin Production (cont.)

Chymosin is an enzyme that has been used in cheese-making for many years. It is produced in the stomach of young mammals and makes milk proteins clot. To make cheese, it used to be necessary to kill animals and extract the chymosin from their stomach. Today, most chymosin is made by GM microbes, usually fungi.

✓ Maximise Your Marks

Be careful when answering questions about insulin. Remember, it is the insulin gene that is put into bacteria, not the insulin molecule. This is a common mistake.

Build Your Understanding

It is possible to use GM bacteria to produce GM plants. The type of bacteria used (*Agrobacterium tumefaciens*) injects its genetic material into plant cells and produces a growth of plant cells called a gall which can be grown into a complete GM plant.

Several different genes can be inserted:
- A gene may make the plant resistant to herbicides. This means the farmer can spray his fields to kill weeds without killing the crop.
- A gene from another type of bacteria (*Bacillus thuringiensis*) can be inserted. This makes the crop plant make its own insecticide.

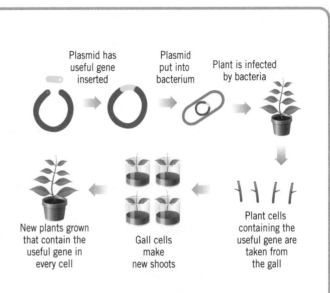

Plasmid has useful gene inserted

Plasmid put into bacterium

Plant is infected by bacteria

New plants grown that contain the useful gene in every cell

Gall cells make new shoots

Plant cells containing the useful gene are taken from the gall

Gene Probes

Using the enzymes involved in genetic engineering, it is possible to identify specific lengths of DNA from a person:
- Step 1 – the regions of the DNA are isolated and cut up using enzymes.
- Step 2 – the DNA fragments are put on a gel and separated using an electric current.
- Step 3 – the fragments are treated with a radioactive or fluorescent probe.

This technique could be used in gene testing to see if a person has a specific harmful allele.

It can also be used to make a **genetic fingerprint**. A genetic fingerprint is a pattern of DNA that can be used to identify an individual.

? Test Yourself

1. What is a transgenic organism?
2. Why should farmers want to make their crops produce insecticide?
3. Write down one advantage of using human insulin rather than pig insulin to treat diabetes.
4. Why can cheese made with GM chymosin be eaten by vegetarians?

★ Stretch Yourself

1. Farmers grow crops that have a gene for resistance to weedkiller inserted in them using *Agrobacterium*. Why do they want to grow these crops?
2. Why is a radioactive or fluorescent probe used in genetic testing?

Microbes

Practice Questions

Complete these exam-style questions to test your understanding. Check your answers on page 126. You may wish to answer these questions on a separate piece of paper.

1 The following processes involve manipulating genes. Match words **A**, **B**, **C** and **D** with numbers **1–4** in the sentences. (4)

A genetic engineering   **B** gene therapy   **C** selective breeding   **D** genetic fingerprinting

Farmers have been manipulating the genes of their animals by choosing which animals to mate.

This is called _____1_____. It is now possible to make bacteria produce human

insulin by the process of _____2_____. If a person's genes are altered to cure a

disease this is called _____3_____. A scientist may use

_____4_____ to identify a person from a sample of their cells.

2 The diagram shows some stages in the production of human insulin by bacteria.

A Insulin gene / Bacterial chromosome
B Bacterium / Chromosome
C
D Human DNA / Insulin gene

a) Write down the order in which the stages take place. The first one has been done for you.

D   ____   ____   ____ (1)

b) The table contains some statements about this method of making insulin. Put a cross (**✗**) next to any incorrect statements. (2)

The process uses hormones to cut DNA.	
The bacteria can be grown in large fermenters.	
The insulin produced is a hybrid of human and bacterial insulin.	
The bacteria can be grown on cheap waste products.	

3 The boxes contain some parts of a fermenter and their jobs. Draw straight lines to join each part to its correct job. (3)

Part
Steam inlet
Water jacket
Stirrer
Air inlet

Job
To keep a constant temperature in the fermenter.
To mix the microorganism with the food.
To allow the microoganism to respire.
To sterilise the fermenter between batches.

4 Look at the diagram, which shows a digester used to make biogas.

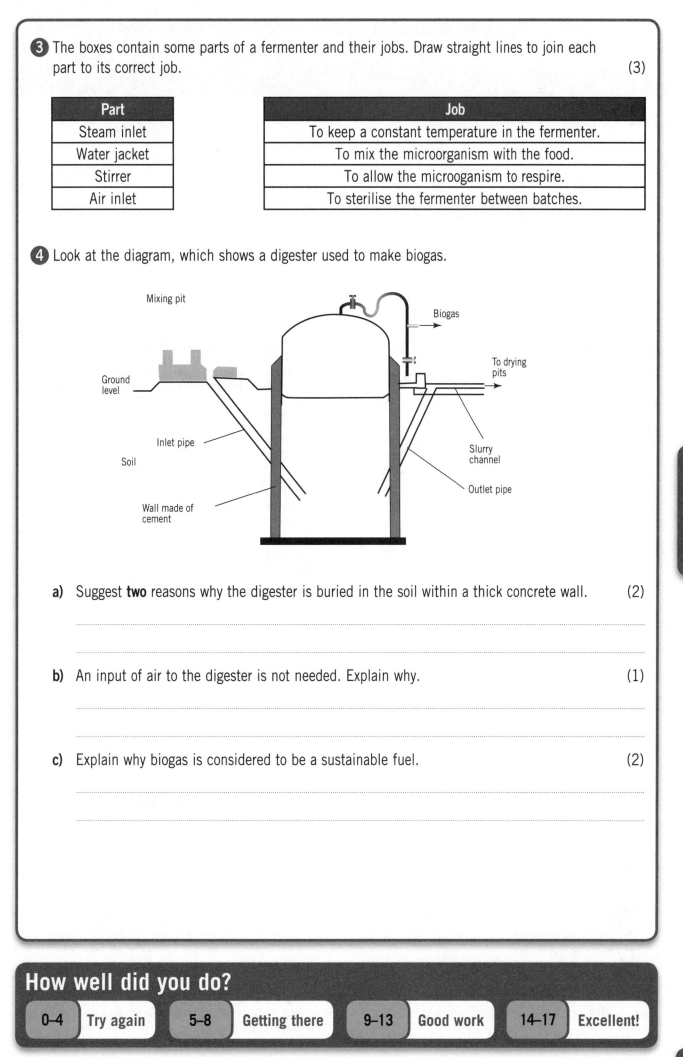

a) Suggest **two** reasons why the digester is buried in the soil within a thick concrete wall. (2)

....

....

b) An input of air to the digester is not needed. Explain why. (1)

....

....

c) Explain why biogas is considered to be a sustainable fuel. (2)

....

....

Microbes

How well did you do?

| 0–4 | Try again | 5–8 | Getting there | 9–13 | Good work | 14–17 | Excellent! |

Answers

Topic 1 Organisms in Action

A Balanced Diet
Test Yourself
1. Amino acids
2. $55 \times 0.6 = 33$ g
3. Kwashiorkor is a deficiency disease caused by a lack of protein in the diet.
4. To move food along the gut (avoid constipation).

Stretch Yourself
1. Egg and beef are first class proteins so contain all the essential amino acids.
2. Peas and wheat are both deficient in a different amino acid so they complement each other.

Homeostasis 1
Test Yourself
1. A hormone is a chemical messenger that causes a response in the body.
2. Insulin is made and released in the pancreas.
3. Insulin causes the liver to convert glucose into the storage carbohydrate glycogen.
4. One test for diabetes is using urine testing sticks to check for glucose.

Stretch Yourself
1. Glucose is a sugar used for respiration in the body, glucagon is a hormone that converts glycogen into glucose and glycogen is a storage carbohydrate.
2. Exercise will reduce the blood sugar level as glucose is used up in respiration.

Homeostasis 2
Test Yourself
1. 37°C
2. We look red due to vasodilation – blood vessels in the skin become wider allowing more blood to the surface so more heat is lost.
3. Too much water is lost in sweating to try and lose heat.
4. The kidneys.

Stretch Yourself
1. Blood is diverted away from extremities to the essential organs, so fingers and toes do not get essential oxygen and glucose for respiration and so die.
2. People who have inactive pituitary glands produce too little ADH and so the kidneys do not reabsorb much water and it passes out in the urine.

Hormones and Reproduction
Test Yourself
1. Testosterone is made in the testes and it causes the development of secondary sexual characteristics in men including the production of sperm.
2. Ovulation is the release of an egg from an ovary.
3. IVF stands for *In Vitro* Fertilisation which is the process by which an egg is artificially fertilised by a sperm outside the body and an embryo re-implanted back into the uterus.
4. Oral contraceptives prevent the release of follicle stimulating hormone (FSH) and so stop ovulation happening.

Stretch Yourself
1. To maintain the uterus lining and to prevent the period (menstruation) happening.
2. The high levels of oestrogen in the pill might produce side effects.

Responding to the Environment
Test Yourself
1. Receptors in the skin can detect touch, pressure, pain and temperature.
2. Glands are another type of effector.
3. The iris controls the size of the pupil/the amount of light entering the eye.
4. A rabbit has eyes on the side of its head so that it can see predators approaching from all angles.

Stretch Yourself
1. Light rays are focusing behind the retina so they need to be bent/refracted more.
2. Due to the ciliary muscles being contracted for a long time.

The Nervous System
Test Yourself
1. A neurone is a single nerve cell, but a nerve contains thousands of neurones.
2. The fatty sheath around a neurone is there to insulate the neurone and speed up the rate of conduction.
3. A synapse is a small gap between two neurones.
4. It is important that reflexes are fast so that they can protect the body from damage.

Stretch Yourself
1. Scientists could design a painkiller that could block the impulse travelling from the pain receptor, e.g. it might act on the synapses in the pathway, preventing the neurotransmitter from passing.
2. It is important that the body breaks down neurotransmitter molecules once they have stimulated a nerve impulse in the next neurone otherwise they would stay in the receptor site and keep sending impulses. This could lead to convulsions.

Plant Responses
Test Yourself
1. Positively phototrophic means plants grow towards light.
2. Positively geotropic means roots grow down into the soil anchoring the plant and finding water.
3. Auxins are found in the shoot tip.
4. Shoots are dipped into hormone powder to make them produce lateral roots.

Stretch Yourself
1. The foil cap blocks the light, so auxin is evenly distributed in the shoot and so the shoot grows upwards without bending.
2. Advantage: more convenient to eat or more flesh to eat; Disadvantage: cannot collect seeds to grow new plants.

Practice Questions
1. A balanced diet contains **seven** main groups of food chemicals. One group is protein which is necessary for **growth**. Proteins are made from molecules called **amino acids**. Minerals are another group of chemicals. An example is **iron** which is needed to make haemoglobin.
2. a) Insulin converts excess glucose into glycogen in the liver.
   b) Blood glucose levels depend on the diet/activity.
   c) From: Don't have to break the skin; less chance of infection; can monitor it continuously; the diabetic does not need to test their own blood and work out how much insulin to administer.

3. a)

	Name of hormone	In which half of the cycle does the hormone reach its highest level?
Hormone A	oestrogen	first half
Hormone B	progesterone	second half

   b) The ovaries.
   c) Largest increase in temperature around time of ovulation; avoid intercourse around the time of the temperature increase.
   d) How the methods work – from: Use injections of FSH; stimulate ovulation; used for women who are not ovulating regularly; eggs could be collected; used for IVF; for women who have blocked fallopian tubes.
   Why some people think that they should not be available free of charge on the National Health Service: Costly procedure; some people say that it is not necessary as the women are not ill and not being able to have children is not a health problem. The person will not physically suffer because of it.

4 Experiment 1: Plastic prevents auxin passing down left side; so right side grows more that left side.
Experiment 2: Auxin produced in the shoot tip; therefore only left side of shoot gets any auxin and grows.

Topic 2 Health and Disease

Pathagons and Infection
Test Yourself
1 Anaemia/scurvy.
2 Protozoa.
3 An antibody is a molecule that attacks pathogens.
4 A phagocyte is a white blood cell that engulfs pathogens.

Stretch Yourself
1 They will mate with females, but no offspring will be produced, therefore less mosquitoes to spread the parasite. They can't reproduce so the population of mosquitoes goes down. Mosquitoes transmit the malarial parasite.
2 Memory cells live for a long time in the body and rapidly produce antibodies if the virus returns. Memory cells are produced that will produce antibodies quickly if the measles virus reenters the body.

Antibiotics and Antiseptics
Test Yourself
1 Flu is caused by a virus. Antibiotics do not destroy viruses.
2 Penicillin.
3 Antiseptics are manmade, but antibiotics are made by microbes. Antiseptics are used outside the body, but antibiotics are used inside.
4 So that microbes that live in the human body are not cultured.

Stretch Yourself
1 Help them to kill other microbes such as bacteria and so avoid competition for food.
2 Microbes might be resistant to certain antibiotics but not to a large number of them.

Vaccinations
Test Yourself
1 A vaccine contains a dead or a weakened form of the pathogen.
2 a) Passive immunity.
   b) Active immunity.
3 In a) the antibodies have been made by a different organism so it is passive, in b) they have been made by the person so it is active.

Stretch Yourself
1 If the vaccine contains live, but weakened, microbes then the person may get some symptoms of the disease.
2 People are not immune to the pathogen anymore, so the disease can re-infect.

Drugs
Test Yourself
1 Drugs such as cocaine are very addictive and changes the body chemistry, so that the body cannot function normally without it.
2 To indicate how dangerous they are and give guidance about punishments for illegal use.
3 To see if they have any side effects/test for safety.
4 They think that it is cruel/they may not have the same effect on animals.

Stretch Yourself
1 Alcohol is a depressant – it will reduce synaptic transmission in neurones involved in pain.
2 So that they can report the results without any bias and do not subconsciously interpret the same results in a different way.

Smoking and Drinking
Test Yourself
1 Increases production of mucus.
2 Walls of alveoli are broken down so reducing surface area for gas exchange; Alveoli are damaged so there is less gas exchange surface.
3 Cirrhosis.
4 Six.

Stretch Yourself
1 So that blood is pumped faster around the body in which the cells receive sufficient oxygen to respire aerobically.
2 15 times more compared with 5 times more – so an increase of three times.

Too Much or Too Little
Test Yourself
1 80/1.72 = 27.7
2 The highest figure is the systolic value (when the heart contracts) and the lowest figure is the diastolic value (when the heart relaxes).
3 A small build up of fat on the inside the walls of the arteries.
4 Less stress/do not drink large amounts of alcohol/regular exercise/less saturated fat in diet/do not smoke.

Stretch Yourself
1 Poor circulation results in limited blood supply to extremities which would usually bring heat.
2 Prevent a thrombosis happening in the coronary arteries which could block them and lead to a heart attack.

Practice Questions
1 An organism that causes a disease is called a **pathogen**. The body tries to prevent these organisms entering the body. Tears contain **lysozyme** and the stomach makes **acid** both of which can destroy the organisms. If the organism enters the body it can damage cells or release poisons called **toxins**.
2 a) Nicotine.
   b) Cannot do without it; lack of the drug causes withdrawal symptoms.
   c) Slow it down; blocks synapses.
   d) Barbiturates are more dangerous; harsher penalties for illegal possession.
3 a) Measles can cause babies to die/mumps can cause deafness in young children.
   b) Because if they catch rubella their babies may become brain damaged.
   c) Three from: contains a weakened or dead pathogen; stimulates the production of antibodies; from white blood cells; memory cells are made; if the live pathogen invades it can be killed rapidly.
4. a) To see if they work/they are safe.
   b) So that a person does not know if they have taken a drug or not; to eliminate psychological effects.
   c) For: animals were suffering unnecessarily; they do not always react in the same way as people.
   Against: this is only one case whereas thousands of drugs have been tested; testing on animals may have saved thousands of lives.
5 a) The more milk drunk, the greater the death rate.
   b) Milk contains saturated fat; this can make it more likely for plaques to form in the coronary arteries; these may block the arteries and cause a heart attack which can be fatal.

Topic 3 Genetics and Evolution

Genes and Chromosomes
Test Yourself
1 Deoxyribonucleic acid.
2 Codes for proteins.
3 So that at fertilisation the full number of chromosomes can be restored.
4 Mutation.

Stretch Yourself
1 Sperm are either X or Y in even numbers. If a Y sperm fertilises an egg then it is a boy and an X sperm makes a girl.
2 Similarities are due to genetics (they have the same alleles) and so any differences must be due to the environment.

Passing on Genes
Test Yourself
1 The allele for rolling is dominant over the allele for non-rolling.
2 Huntington's disease.
3 Genotype is what alleles a person has and phenotype is how the alleles express themselves (the characteristics of the person).
4 Testing for a genetic disease.

Stretch Yourself
1 A gene is a length of DNA that codes for a protein, an allele is a particular copy of a gene that codes for a particular variation of the protein.
2 They may not get insurance or the job they apply for if people know that they will develop a genetic disease in the future.

Introduction to Gene Technology
Test Yourself
1 Cuttings will produce an identical copy so the gardener can be sure of the characteristic of the plant.
2 Dolly was a sheep and the first mammal that was produced by cloning from an adult cell.

3 They can differentiate into any other type of cell.
4 Resistance to insects eating them/resistance to herbicides/produce a higher yield.

Stretch Yourself
1 Use of embryonic stem cells involves the destruction of an embryo, but using adult stem cells doesn't. Some people consider an embryo to be an individual life.
2 They can spray their whole field with weedkiller, killing all the weeds, except the crop.

Evolution and Natural Selection
Test Yourself
1 The word extinct means that all the individuals of a species no longer exist.
2 Any dinosaur species.
3 They were camouflaged against the polluted buildings and trees, so less likely to be caught by predators.
4 Darwin thought that religious people would be against his ideas because his theory went against the beliefs at the time.

Stretch Yourself
1 All giraffes show variation in neck length. The giraffes with longest necks can reach the leaves and are more likely to survive. They are more likely to reproduce and pass on the genes for long necks. Over time this leads to the giraffe population having longer and longer necks.
2 The best adapted organisms are the ones that are most likely to survive in an environment.

Practice Questions
1 C
2 D
3 A
4 a) The genes in cells are found in the part of the cell called the **nucleus**.
  b) These genes are on long strands called **chromosomes**.
  c) They are made of a chemical called **DNA**.
  d) The genes control the cell by describing which **proteins** the cell should make.
5 a) There would only be 23; they would not be in pairs.
  b) The last pair are different – it contains an X and a Y chromosome.
6 a) Recessive; Jackie and Leroy have the allele but not the disorder.
  b)

gametes	F	f
F	FF	Ff
f	Ff	ff

  c) 25%
  d) If the baby did have cystic fibrosis, Jackie would have to decide whether to have an abortion or not.
7 Some rats had an allele for resistance.
  They survived the poisoning and reproduced more than those without the allele.
  So they passed the resistance allele to their offspring and overtime the population became more and more resistant.

Topic 4 Organisms and Environment

Classifying Organisms
Test Yourself
1 A fungus.
2 The offspring of a cross between members of two closely related species.
3 No, because they are not in the same species.
4 They have fur and produce milk.

Stretch Yourself
1 They are both adapted to living in the same conditions, i.e. water.
2 Because organisms have evolved from common ancestors by a gradual process there are always going to be organisms that have characteristics that are intermediate between groups.

Competition and Adaptation
Test Yourself
1 Light is needed it for photosynthesis.

2 Thick fur on top insulates them from the suns heat, thin fur underneath allows heat to escape.
3 They are reduced to spines so that less water is lost.
4 They have eyes on the side of their heads so that they have virtually all round vision.

Stretch Yourself
1 Many animals that live near the poles give birth to live young and often migrate away from the pole in the winter. Most animals in cold regions give birth to live young because eggs are likely to get too cold and the foetus dies.
2 Large animals have a small surface area to volume ratio and so could overheat. Elephants have big ears to increase their surface area and lose excess heat.

Living Together
Test Yourself
1 Predators have more food, so are able to reproduce more.
2 They live on a living organism and take food from them so cause them harm.
3 It can make some of its own food by photosynthesis, but also takes some resources from the tree it grows on.
4 To attract and reward insects so they pollinate them.

Stretch Yourself
1 Fungi cannot photosynthesise, so they get food from the algae.
2 Very little sunlight can penetrate deep in the oceans, so there are no plants as photosynthesis cannot occur.

Energy Flow
Test Yourself
1 Many organisms eat more than one type of food.
2 They might decrease because the birds would eat more of them as there are no blackflys to eat.
3 The amount of biomass of living material at each trophic level.
4 Heat/excretion/egestion/uneaten parts from respiration.

Stretch Yourself
1 $4/20 \times 100 = 20\%$
2 This is much more efficient. This is because meat contains more energy and it is easier to digest than plant matter.

Recycling
Test Yourself
1 To allow the microbes to decompose the plant material – decomposers need water to be able to survive and decompose.
2 Respiration/combustion.
3 Nitrates.
4 To make proteins.

Stretch Yourself
1 If the buildings are made of limestone then the acid will react with the carbonate in the limestone releasing carbon dioxide.
2 This will allow decomposers in the soil to release minerals. It will also allow nitrifying bacteria to produce nitrates.

Populations and Pollution
Test Yourself
1 Exponential means that the rate of increase is increasing.
2 Carbon dioxide/methane.
3 Sulfur dioxide.
4 Rat-tailed maggots.

Stretch Yourself
1 Transportation of that food to us involves the burning of fossil fuels that release carbon dioxide.
2 Sewage contains nitrogen containing compounds, so can also cause an algal bloom as the algae use it for growth.

Conservation and Sustainability
Test Yourself
1 Overhunting/destroying habitats/causing pollution/changing of climate.
2 The variety of different organisms in a habitat.
3 Trying to preserve a habitat so that the organisms that live there are protected.
4 Fewer resources are used to make new materials and there is less waste.

Stretch Yourself [S/H]

1 The small patches may only contain small numbers of individuals of a species. There might not be enough genetic diversity.
2 Cheetahs do not have much genetic diversity as they are all related to a small number of individuals. They try and reproduce them with other cheetahs that are not closely related and so have less similarities in their genes. Using artificial insemination from less related individuals will help reduce inbreeding problems.

Practice Questions

1 C
2 B
3 Amphibian – Has moist permeable skin.
Fish – Has wet scales.
Invertebrate – Does not have a backbone.
Mammal – Is covered in fur.
Reptile – Has dry scales.
4 a) The hump of a camel can be used to provide energy when food is scarce; do not need to eat so often/helps insulate the body from the sun.
  b) Deep roots can absorb water from deep underground can get water from different parts of the soil; roots that are spread out can absorb water over a wide area; roots that are shallow absorb the water before it evaporates.
  c) Large animals cool down more slowly; decreases surface area to volume ratio so less heat loss; small ears reduce the surface area.
  d) Less competition between the larvae and adults.
5 a) Ladybirds eat greenfly; greenfly increased in July because plenty of food/warmth; more food for ladybirds so their numbers increased; greenfly numbers dropped because more were eaten; so ladybird numbers started to drop. This pattern was repeated.
  b) Predator–prey graph.
6 a) Their habitat (i.e. the sea) is so vast and they live deep under the sea so that it is impossible to estimate population size.
  b) For: they have quotas so at least there is a limit to the numbers caught only hunt for scientific research; vary the type of whales being hunted. Against: have caught fin whales that are protected; some people are not convinced that it is for scientific research;

Topic 5 Cells and Organisation

Cells and Organisation
Test Yourself

1 Cell wall, vacuole, chloroplasts.
2 Cell sap can have many functions. It helps support the plant.
3 Organs.
4 In the cytoplasm.

Stretch Yourself

1 10 mm wide so 200 times (10 mm / 0.005 mm = 200 mm)
2 The cells become so specialised that they cannot take on other roles and take the function of those lost.

DNA and Protein Synthesis
Test Yourself

1 Because two helices are twisted into a spiral shape.
2 The structure has a different order of amino acids.
3 Hydrogen bonds between the bases, C with G and A with T.
4 Each sequence of three bases codes for one amino acid.

Stretch Yourself

1 16%
2 The double helix unwinds and the base pairs separate. This allows mRNA bases to pair with the DNA bases producing a complimentary copy to pass to the ribosomes.

Proteins and Enzymes
Test Yourself

1 Proteins are needed to make key structures inside the body, e.g. bone and so without protein growth will be limited.
2 To allow reactions to be fast enough at body temperature.
3 The lock is the enzymes active site and the key is the substrate.
4 The temperature at which the reaction occurs at the fastest rate.

Stretch Yourself

1 Lipase has a particular shaped active site that fats will fit into, but not proteins.
2 Vinegar is acidic and so the pH would be too low so the enzymes of the decay organisms would not work.

Cell Division
Test Yourself

1 One
2 In the ovaries and testes.
3 46
4 Meiosis introduces more variation; meiosis makes four cells but mitosis makes two; meiosis makes cells with half the number of chromosomes, but mitosis produces cells that have the same number as the parent cells. Mitosis produces cells that are genetically identical to each other and to the parents – meiosis produces genetically different cells.

Stretch Yourself

1 It usually produces a new protein that does not work as well.
2 UV light can be absorbed by DNA and cause mutations.

Growth and Development
Test Yourself

1 Muscle cell.
2 A cell that has not yet differentiated and so can divide to produce any type of cell.
3 Growth becomes negative, i.e. more cells are dying than are being produced.
4 The meristems at the tips of the roots and shoots.

Stretch Yourself

1 The brain needs to develop first to control the other parts of the body.
2 This is destructive and it is difficult to tell when it has all been removed; Organism has to be killed and all water removed in order to measure dry mass. Therefore it is hard to get an idea of growth over time.

Transport in Cells
Test Yourself

1 The molecules from the 'stink bomb' move from a high concentration to a low concentration over the other side of the room. It diffuses.
2 Diffusion is faster in warm conditions because the molecules are moving faster.
3 A membrane that lets some molecules through but not others.
4 Water

Stretch Yourself

1 0.16 moles
2 The air spaces between the soil particles have been filled with water which takes the place of oxygen.

Respiration
Test Yourself

1 We need to respire more to generate more heat to keep a constant body temperature, so more food is needed to provide glucose for respiration to produce this heat.
2 Respiration rate increases to supply extra energy for muscle contraction. More oxygen is therefore needed and more carbon dioxide needs to be removed.
3 Carbon dioxide.
4 Lactic acid is produced due to anaerobic respiration.

Stretch Yourself

1 6/6 = 1.0
2 So that the temperature is at an optimum level for the yeast enzymes to produce alcohol.

Practice Questions

1 All organisms release energy from food, this largely happens in the **mitochondria**. Cells take up water by osmosis because the **cell membrane** is partially permeable. The **vacuole** stores some sugars and salts. Plant cells are limited to how much water they can take up because the **cell wall** resists the uptake of too much water.

Answers

123

2

	Osmosis	Diffusion	Active transport
Can cause a substance to enter a cell.	✓	✓	✓
Needs energy from respiration.	✗	✗	✓
Can move a substance against a concentration gradient.	✗	✗	✓
Is responsible for oxygen moving into the red blood cells in the lungs.	✗	✓	✗

3 Amino acids – These chemicals join together to make a protein molecule.
   Bases – The order of these chemicals on the chromosomes codes for proteins.
   DNA – Chromosomes are made from this chemical.
   RNA – This is the chemical messenger that carries the genetic code out of the nucleus.
4 a) On the chromosomes; in the nucleus.
   b) Each gene has a different order of DNA. The bases code for the order of amino acids in the protein.
   c) 1094/20 000 × 100 = 5.47%
   d) The liver – it uses the most genes and they code for different enzymes.
5 Stem cells are undifferentiated cells, i.e. they have the potential to become any type of cell. They may be used to replace genetically diseased cells to cure genetic diseases. Some people object to the use of embryos to provide stem cells as it destroys the embryo. If this process works then embryonic stems cells would not have to be used.

Topic 6 Ecology

Sampling Organisms
Test Yourself
1 A population.
2 a) quadrat b) net c) pitfall trap.
3 20 in 1m² so 2000 in the field.
4 One area of the field may not be representative of all the areas. It saves time.

Stretch Yourself
1 210
2 Some animals can survive more time out of the water than others and so can live further up the shore.

Photosynthesis
Test Yourself
1 To absorb the energy of sunlight.
2 Palisade mesophyll.
3 Stomata are structures containing leaf pores that allow gases in and out of the leaf.
4 Leaves have veins to supply water for photosynthesis and to take away the sugars that are produced from photosynthesis.

Stretch Yourself
1 It is released from water.
2 High temperatures will denature the enzymes that control the reactions of photosynthesis.

Food Production
Test Yourself
1 Plants need nitrates to produce amino acids and therefore proteins.
2 Chlorophyll cannot be made, as chlorophyll contains magnesium.
3 Growing plants without the use of soil just in water containing nutrients.
4 The temperature is not low enough to completely stop the growth of microbes that cause decay.

Stretch Yourself
1 Often they kill the natural predators of the pests too and so increase the problem. They also contaminate the environment and reduce biodiversity.
2 It prevents loss of energy in pig movement therefore leaving more energy for growth.

Practice Questions
1

Spreading manure on the fields.	
Spraying chemical pesticides.	✗
Killing weeds using weedkillers.	✗
Rotating their crops.	

2 a) Pondweed lives underwater, makes bubbles of oxygen that can be counted, it is easy to obtain and photosynthesises readily.
   b) Oxygen.
   c) He uses the same piece of pondweed throughout.
   d) Use a more accurate timer, such as a stop watch.
3 a) Stretch a tape measure down the shore. Use a quadrat to sample at different distances down the shore.
   b) Half way down the shore. Spread between quadrats 3 and 9 – most are in quadrat 7.
   c) Further up the shore the seaweed might dry out; too far down the shore then it might be covered too much; and not get enough light.
4 a) More likely to believe it; tend to trust scientists/assume a report must be well researched.
   b) The text does not give a balanced argument. It concentrates on the environmental issues and not health issues.
   c) Some countries struggle to produce enough food; yields might drop if the food is grown organically.

Topic 7 Physiology

Transport in Animals
Test Yourself
1 Platelets clot the blood.
2 To fit more haemoglobin in, so they can carry more oxygen.
3 To stop the blood flowing backwards as the pressure is low.
4 Pulmonary artery.

Stretch Yourself
1 It contains deoxygenated blood (which is dark red, not blue).
2 When the left ventricle contracts some of the blood goes back into the left atrium rather than out into the aorta. This causes a backlog of blood in the veins coming back from the lungs.

Transport in Plants
Test Yourself
1 In a plant stem water and minerals move upwards, towards the leaves.
2 The movement of dissolved food through the phloem.
3 Running down the centre.
4 To increase the surface area for water absorption.

Stretch Yourself
1 Because carbon dioxide must be allowed in for photosynthesis.
2 The guard cells lose water by osmosis and so the cells become flaccid, straightening up and closing the pore.

Digestion and Absorption
Test Yourself
1 The starch is being digested into the sugar maltose by amylase.
2 The gall bladder stores bile.
3 Fatty acids and glycerol.
4 It is sweeter so less is needed.

Stretch Yourself
1 So that the acid does not damage it and the protease does not digest it.
2 Slows down the rate of absorption as there is a smaller surface area.

Practice Questions
1 Plants take water up from the soil. Plants have many **root hairs** to increase their surface area for water uptake. The water is carried up the stem in the **xylem**. Sugars, however, are transported in the **phloem**. Water is lost to the air through **stomata**.
2 Aorta – Carries deoxygenated blood away from the heart.
   Pulmonary artery – Carries oxygenated blood under low pressure.
   Pulmonary vein – Carries blood into the right atrium.
   Vena cava – Carries oxygenated blood under high pressure.
3 Starch molecules are too large to be able to pass into the bloodstream and so need to be **digested** first. This digestion begins in the **mouth**. An enzyme called **amylase** breaks down starch into maltose. Maltose is then

digested into **glucose** in the small intestine. Absorption then occurs and this is speeded up by the presence of tiny projections on the wall of the small intestine called **villi**.

4 a) A = artery; B = vein; C = capillary
   b) B
   c) Wall is one cell thick; allows exchange of substances with the tissues.
5 a) To prevent air bubbles entering the xylem vessels.
   b) Water evaporates from the leaves – this draws water up the xylem and from the tube.
   c) Fan will increase the rate of movement – this is because water vapour will be blown away, allowing more to evaporate.
   Petroleum jelly will decrease the rate of movement – this is because the jelly will block the stomata.

Topic 8 Use, Damage and Repair

The Heart and Circulation
Test Yourself
1 A single, closed circulation.
2 The microscopes at the time were not powerful enough.
3 Respiration.
4 To prevent blood clotting in blood vessels.

Stretch Yourself
1 Blood is at higher pressure and moves in pulses rather than smooth. In the arteries, blood is under high pressure and moves in pulses. In veins, it is under lower pressure and moves more evenly. The rate of flow is the same in both.
2 Anti A antibodies and rhesus antibodies, i.e. Anti B and Rh+

The Skeleton and Exercise
Test Yourself
1 Because they contain synovial fluid they allow smooth movement.
2 So that the bird can fly easily using as little energy as possible.
3 A ball and socket joint.
4 Tendons join muscle to bone and ligaments join bone to bone and are more elastic.

Stretch Yourself
1 When it contracts, the arm is extended or straightened.
2 They get through the filter but are then selectively re-absorbed back into the bloodstream.

The Excretory System
Test Yourself
1 Water, salts and urea.
2 A certain blood pressure is needed to filter the blood.
3 Protein molecules are too large to get through the filter.
4 They are filtered out of the blood but then are reabsorbed.

Stretch Yourself
1 Less water is reabsorbed so a greater volume of urine is produced.
2 The protein is digested into amino acids and the excess amino acids are broken down into urea which is excreted in the urine.

Breathing
Test Yourself
1 By diffusion over its body surface, through its skin.
2 Because it is moist enough for them to breathe through their skin.
3 In the filaments.
4 It contracts and flattens.

Stretch Yourself [S/H]
1 Because they have a rich blood supply.
2 About 4500 cm^3.

Damage and Repair
Test Yourself
1 Because the blood on the left is oxygenated and the right is deoxygenated.
2 Urea/excess water/excess salts.
3 Osteoporosis makes the bones weaker.
4 It is inherited; a combination of two recessive cystic fibrosis alleles.

Stretch Yourself
1 So that they do not pass out of the blood.
2 They relax the smooth muscle so allowing the bronchioles to dilate.

Transplants and Donations
Test Yourself
1 Genetically more similar so less likely for rejection to occur.
2 The battery might need to be changed.
3 Light/does not corrode/smooth.
4 The surrogate may not want to hand over the baby; the baby is not genetically linked to the mother.

Stretch Yourself
1 People with B- blood have anti-A antibodies in their blood and so would attack the red blood cells causing agglutination.
2 They can give their blood to people of any other blood group as their blood contains no anti A or B antigens.

Practice Questions
1 A condition that weakens the bones is called **osteoporosis**. Coronary heart disease may be treated by a **bypass**. The function of the kidneys may be replaced by **dialysis**. A woman who cannot maintain a pregnancy may get a baby using **surrogacy**.
2

	glucose	protein	urea
Present in the blood reaching the kidney.	✓	✓	✓
Passes out of the blood in the filter unit.	✓	✗	✓
Re-absorbed back into the blood from the kidney tubule.	✓	✗	✗
Usually present in the urine.	✗	✗	✓

3 Bone – Living tissue containing cells and calcium salts.
   Cartilage – Shiny substance that reduces friction in joints.
   Ligaments – Elastic structure that holds joints together.
   Tendons – Inelastic structure that joins muscle to bone.
4 a) Much thicker muscle wall in left ventricle. It needs to pump the blood further.
   b) Pressure is less in right ventricle as blood is only pumped to lungs and not to the whole body
5 a) Alveoli.
   b) Pollen, dust, animal fur.
   c) Vital capacity; Tidal volume.
   d) Smaller tidal volume – about 1 litre compared to 1.25.
   Smaller vital capacity – about 3.5 litres compared to 4.
6 There are more on the waiting list than there are transplants. The difference between the number of transplants and the number on the weighting list is increased. The number of transplants is less than the number of donors. Some people donate more than one organ.

Topic 9 Animal Behaviour

Types of Behaviour
Test Yourself
1 So that they stay with their mother for food/protection/to learn.
2 So that they become habituated them to loud noises and do not respond to them when working.
3 The birds become habituated and are not scared – it becomes not associated with danger.
4 The cerebrum/cerebral cortex.

Stretch Yourself
1 Give them a reward every time they do the trick.
2 The size of the brain that they contained can be worked out which gives an idea of intelligence.

Communication and Mating
Test Yourself
1 So that it can compete for females and show that it is healthy and has good genes.
2 They pair for life.
3 It uses up food reserves and makes the parents more vulnerable to predators.
4 Airborne chemical messengers that are produced to attract a mate.

Stretch Yourself

1 All apes have opposable fingers.
2 She lived with them and spent so much time with them they learned to ignore her; She was able to live with gorillas without them becoming alarmed to her presence.

Practice Questions

1 Apes have the ability to communicate using expressions, this is an example of **body language**. They can also communicate using chemicals called **pheromones**. It has been shown that apes can change their behaviour as a result of experiences. This is called **learning**.
2 Courtship – Male peacocks often display their tail feathers.
   Habituation – Police dogs have been trained not to be frightened of loud noises.
   Imprinting – Baby chicks will follow the first moving object that they see.
   Conditioning – Car drivers usually brake when they see a red traffic light without thinking about it.
3 a) It allows a female to chose a male that has good genes, indicates a readiness to mate and it helps with bonding.
   b) Ear tufts are bigger or standing up, necks are longer and bodies facing each other.
   c) To use to build nest, to form a bond and to practice feeding young.

4 a) The dog produced saliva. The dog associated the ticking noise with food and therefore showed conditioned behaviour.
   b) For: behaviour can be observed that may also apply to humans, easier to do than on humans; they may become useful in the future for treating disorders.
      Against: animals are not leading a completely natural life; that the experiment was cruel and infringe animal's rights.
5 Four from: Chimp using learned behaviour; learning by its experiences; it is insight behaviour; new neutral pathways are being set up in the brain; stores memories of the event.

Topic 10 Microbes

The Variety Of Microbes
Test Yourself
1 A small part of the yeast breaks off the parental cell.
2 A long whip-like projection from bacterial cells that allows them to move.
3 Yeast, bacterium, virus.
4 It is produced by the fungus *Penicillium*.

Stretch Yourself
1 The light and temperature is increasing and there are plenty of minerals available.
2 The numbers of phytoplankton have peaked and so there is plenty of food for them.

Putting Microbes to Use
Test Yourself
1 So that any pathogens are destroyed.
2 High temperatures would kill the culture.
3 Because it is made from fungi not animals and is high in protein.
4 Because the alcohol kills the yeast when it reaches about 15%.

Stretch Yourself
1 The trees are only burnt at the same rate that they grow so there is no net carbon dioxide release.
2 It is a habitat containing many rare species / it may contribute to global warming.

Microbes and Genetic Engineering
Test Yourself
1 An organism that contains genes from another species.
2 So that any insect pests will be killed if they try and eat the plants.
3 Works better/has less side effects.
4 It does not involve killing animals to extract the chymosin from their stomachs.

Stretch Yourself
1 They can then spray their crops with weed killer and only the weeds will die, therefore reducing competition.
2 So that the sections of DNA can be seen/photographed; To show where the gene being tested for is.

Practice Questions

1 Farmers have been manipulating the genes of their animals by choosing which animals mate. This is called **selective breeding**. It is now possible to make bacteria produce human insulin by the process of **genetic engineering**. If a person's genes are altered to cure a disease this is called **gene therapy**. A scientist may use **genetic fingerprinting** to identify a person from a sample of their cells.
2 a) D, A, B, C
   b)

The process uses hormones to cut DNA.	✗
The bacteria can be grown in large fermenters.	
The insulin produced is a hybrid of human and bacterial insulin.	✗
The bacteria can be grown on cheap waste products.	

3 Steam inlet – To sterilise the fermenter between batches.
   Water jacket – To keep a constant temperature in the fermenter.
   Stirrer – To mix the microorganism with the food.
   Air inlet – To allow the microrganism to respire.
4 a) Two from: In case of leaks; to keep it warm; to contain any explosion; to eliminate smells.
   b) The microbes are respiring anaerobically.
   c) It gives out carbon dioxide that is taken in by plants growth; no net increase in carbon dioxide concentrations.

Index

Index